A NATURALIST'S GUIDE TO THE

FLOWERS
OF
SRI LANKA

T0206553

Darshani Singhalage, Nadeera Weerasinghe & Gehan de Silva Wijeyeratne

JOHN BEAUFOY PUBLISHING

Reprinted in 2024

First published in the United Kingdom in 2018 by John Beaufoy Publishing Ltd
11 Blenheim Court, 316 Woodstock Road, Oxford OX2 7NS, England
www.johnbeaufoy.com

Photo Credits
Front cover: *main image* Climbing Flax © Gehan de Silva Wijeyeratne, *bottom left*: Sweet Granadilla © Nadeera
Weerasinghe, *bottom centre*: Butterfly Pea © Nadeera Weerasinghe *bottom right* Marsh Henna © Nadeera Weerasinghe
Back cover: Crown Flower © Gehan de Silva Wijeyeratne
Title page: Snake Thumb © Nadeera Weerasinghe
Contents page: Small-leaf Osbeckia © Nadeera Weerasinghe
Main descriptions: Photographs are denoted by a page number followed by t (top), b (bottom), l (left), c
(centre) or r (right).
Gehan de Silva Wijeyeratne 26bl, 27b, 35tr, 39br, 41br, 44bl, 49tr, 63br, 67bl, 67br, 68b, 69bl, 69br, 70bl, 75 (all
three), 77 (both), 78b, 80bl, 83tc, 90tl, 91br, 93bl, 97 (both), 100tl, 100br, 102bl, 103tr, 107tl, 107tr, 108tl, 112
(both), 113 (both), 115tr, 117b, 118tl, 118br, 121 (both), 128 (both), 129tr, 129br, 131b, 133cr, 133bl, 140bl,
142tl, 142tr, 143 (both), 146br, 149b, 150bl, 151tr, 151bl, 154bl, 166tl. **Chamara Irugalratne** 42 bl. **Nilantha
Kodithuwakku** 49br, 130bl. **Kavithma Nayanadini** 105tr. **Darshani Singhalage** 31tr, 32br, 36tr, 36tl, 51br, 70br,
124bl, 124br, 135b. **Nadeera Weerasinghe** 21b, 22(all three), 23b, 24 (both), 25b, 26br, 28b(both), 29tr, 30tr, 30tl,
30br, 31br, 32tl, 32bl, 33bl, 33br, 34tr, 34tl, 34br, 34bl, 35br, 35bl, 36bl, 37tr, 37bl, 37br, 38tl, 38bl, 39tr, 40 (both),
41tr, 41bl, 42tl, 43b, 44tl, 44br, 45 (both), 46 (both), 47 (both), 48b, 50 (both), 51tr, 51bl, 52b, 53 (both), 54b,
55(both), 56b, 57b, 58(both), 59(both), 60b, 61(all three), 62b, 63tl, 63tr, 64b, 65tr, 65bl, 65br, 66(both), 67tl, 67tr,
70tl, 71b, 72(all three), 73tr, 73b(all three), 74 (all three), 76 (both), 78tr, 79tr, 79br, 80tl, 80tr, 81tl, 81tr, 81tl, 81tr,
82tl, 82bl, 83tl, 83tr, 83bl, 83br, 84tl, 84tr, 84bl, 84br, 85tr, 85br, 86 (all three), 87 (all three), 88 (all four), 89tr, 89br,
90bl, 91tl, 91tr, 91bl, 91bc, 92 (all three), 93tl, 93tr, 93br, 94(both), 95b, 96tl, 96bl, 98b, 99 (both), 100tr,100bl,101
(both), 103tc, 103br, 104tl, 104bl, 105br, 106 (both), 107bl, 107br, 108tr, 108bl, 108br, 109 (both), 110b, 111 (both),
114tl, 114bl, 115br, 116b, 118tr, 118bl, 119 (both), 120tl, 120bl, 122 (all three), 123 (both), 124tl, 125 (both), 126tl,
126bl, 127 (all four), 129bl, 130tl, 131tr, 132 (all four), 133tl, 133tr, 133br, 134 (all three), 135tl, 135tr, 136b, 137b,
138 (all three), 139 (all three), 140tl, 140tr, 140bc, 140br, 141 (all four), 142b (all three), 144 (both), 145 (both),
146tl, 146bl, 147b, 148b, 151tl, 152 (both), 153 (both), 154br, 155 (both), 156 (all four), 157 (all three), 158 (both),
159 (both), 160 (all four), 161 (both), 162 (all five), 163 (all three), 164 (all four), 165 (all three), 166bl, 166br.

ISBN 978-1-912081-55-4

Edited by Krystyna Mayer

Designed by Gulmohur Press, New Delhi

Printed and bound in Malaysia by Times Offset (M) Sdn. Bhd.

·CONTENTS·

Introduction

This compact book focuses on field use for beginners and experts alike, with data on identifying each plant species, and on its distribution, habitats and flowering period. There is information on the taxonomy and biology of plants, and on sites to visit for key plants, as well as lists of useful references for those who wish to progress further in learning about the beautiful and photogenic flowers of Sri Lanka. Where available, Sinhalese (S) and Tamil (T) names for the featured plants are given in parentheses after the main species headings.

Flowering Plant Species in Sri Lanka

This book does not contain an up-to-date checklist of flowering plant species as that would require a publication in itself, given the large number of species that occur in Sri Lanka. In *A Checklist of the Flowering Plants of Sri Lanka* by Lilamani Kumudini Senaratne, published in 2001, the checklist included 214 families, 1,522 genera and 4,143 species. Of these, about 75 per cent were considered indigenous and about 25 per cent exotics. Of the indigenous plant species, 27.5 per cent were endemic to Sri Lanka. Of the exotics, about a third were naturalized, with the remaining two-thirds being under cultivation. In the two decades since publication, a few more species have been described to science, and the number of exotics that has been introduced has almost certainly increased. However, the figures from 2001 are broadly indicative of the plant species diversity in Sri Lanka.

The list of plants (and hence families) recorded in Sri Lanka continues to grow. There are now 217 families, compared to the 214 families mentioned above. The list has been added to as a result of the introductions of alien plants that have become naturalized, as well as the discovery of indigenous plants that have not been recorded before.

Climatic Zones & Monsoons

The topography of Sri Lanka comprises lowlands along the perimeter, which in the southern half give rise within a short distance to the central hills, rising to above 2,400m in altitude. A careful examination of the topography reveals that the island can be divided into three peneplains, or steps, which were first described by the Canadian geologist Frank Dawson Adams in 1929. The lowest peneplain is from 0 to 30m, the second rises to 480m, and the third to 1,800m. Sri Lanka can be broadly divided into four regions (low-country wet zone, hill zone, low-country dry zone and intermediate zone), resulting from the interactions of rainfall and topography. Rainfall is affected by monsoonal changes that bring rain during two monsoons: the south-west monsoon (May–August) and the north-east monsoon (October–January). Their precipitation is heavily influenced by the central hills. These monsoons deposit rain across Sri Lanka and contribute to the demarcation of climatic regimes.

LOW-COUNTRY WET ZONE

The humid, lowland wet zone in the south-west of Sri Lanka does not have marked seasons, being fed by both the south-west and north-east monsoons. It receives 200–500cm of rain from the south-west monsoon, and afternoon showers from the north-east monsoon. Humidity is high, rarely dropping below 97 per cent, while temperatures range from 27 to 31° C over the year. This zone is the most densely populated area in Sri Lanka.

The coast is well settled, while the interior has Coconut *Cocos nucifera* and Rubber *Hevea* sp. plantations, some rice (paddy) cultivation and small industries. Remnants of rainforests and tropical moist forests exist precariously in some parts of the interior, under pressure from an expanding population. It is in these forests that most of the endemic species that are a draw to ecotourists can be found.

HILL ZONE

The mountainous interior lies within the wet zone and rises to more than 2,400m. Rainfall is generally well distributed, except in the Uva Province, which gets very little rainfall in June–September. Temperatures are cooler than in the lowlands and can vary from chilly in the mornings to warm by noon. In the mid-elevations, such as the area around Kandy, the temperature varies between 17 and 31° C during the year. Temperature variations during a 24-hour cycle are, however, far less varied. The mountains are cooler, within a band of 14–32° C during the year. There may be frost in the higher hills in December and January, when night-time temperatures fall below zero. The central hill zone is intensely planted with tea, but has small areas of remnant forest and open grassland.

LOW-COUNTRY DRY ZONE

The rest of the country, three-quarters of Sri Lanka's land area, consists of the dry zone of the northern, southern and eastern plains. These regions receive 60–190cm of rain each year, mainly from the north-east monsoon. The dry zone is further divided into the arid zones of the north-west and south-east. These areas receive less than 60cm of rain as they are not in the direct path of the monsoonal rains.

The coastal plains in the Southern Province, where Yala and Bundala National Parks are located, and the North Central Province, where the cultural sites are situated, are dry and hot. Much of this zone is under rice and other field crops, irrigated by vast, man-made lakes (tanks, or wewas), many of which are centuries old, and were built by royal decree to capture the scarce rainfall in these areas. Once the 'Granary of the East', exporting rice as far as China and Burma, wars, invasions, and malaria and other diseases laid waste to vast areas of this zone. The once-bountiful rice plains were reclaimed by scrub jungle, the haunt of elephants, bears and Leopards. Since independence in 1948, successive governments have vigorously promoted colonization and

Bundala

resettlement of these areas. Sandy beaches fringe the coastline, and it is always possible to find a beach that is away from the path of the prevalent monsoon.

INTERMEDIATE ZONE

This is a transition zone between the dry and wet zones. Recent rainfall data shows that the wet zone with the highest precipitation is smaller than shown in maps of a few decades ago.

Habitats & Top Sites

For a moderately sized island, Sri Lanka offers a variety of habitats. It is a continental island with a continental shelf coming close to it in the south at Dondra and also near the Kalpitiya Peninsula. Submarine canyons cut into the Trincomalee harbour in the north-east. A chain of partially submerged islands (Adam's Bridge) connects it to the Indian mainland. The entire coastline offers suitable habitat for mangroves, with the western coast especially off Kalpitiya and Mannar being quite rich in these plants. Sri Lanka is dotted with more than 2,000 man-made freshwater lakes in the dry zone. As many as 103 river systems create a rich network of aquatic habitats that are further enriched by paddy fields, which form an important artificial system of wetlands. The dry zone is characterized by grassland, thorn scrub and wooded sections where the soil and rainfall support them, or where 'gallery forests' remain along watercourses. The lowlands in the south-west originally held rainforests, though much of these have been lost to cultivation. At higher elevations, the highlands in the wet zone hold cloud forests, now a tiny remnant of what once existed. The highlands are interspersed with patana grassland.

In the context of visiting foreign botanists, the endemics that are found in the wet zone would be key targets. For this, a good lowland rainforest site such as Sinharaja and a montane site like Horton Plains National Park are essential places to visit. Another accessible rainforest is Kithulgala Rainforest. Given that Sri Lanka is the best place for big-game safaris outside Africa, a visit to a national park such as Yala is also recommended, which allows a chance to see some of the plants in the dry zone. For general information on wildlife highlights and a wildlife-viewing calendar, *Wild Sri Lanka*, published by John Beaufoy Publishing, is recommended. A PDF of the article 'Why Sri Lanka is Super-rich for Wildlife' by Gehan de Silva Wijeyeratne is also available on the Internet.

Botanizing is, however, more complex than going in search of birds and the larger mammals, which can be taken in by visiting a few key sites. Sri Lanka has a number of floristic regions and many species are confined to small areas. There are also species confined to sites such as Ritigala, an inselberg in the dry zone of the northern plains, and areas such as Nilgala or Moneragala in the east. Isolated mountain wildernesses such as the Knuckles hold endemic species, and there are probably many species still awaiting discovery. Even well-known parks and reserves such as Horton Plains and Sinharaja have had attention from a limited number of botanists and probably hold species unknown to science.

Listed opposite are the key sites visited on tours for birds and other wildlife, which are just as good for observing plants.

KEY SITES

Lowland Wet Zone
Talangama Wetland In Colombo's suburbs, visited on arrival or before departure. Good range of wetland plants and trees in urban environments.
Bodhinagala Small but rich patch of rainforest about an hour and a half from Colombo.
Sinharaja The most important site for Sri Lanka's endemic fauna. Half the trees here are endemic.

Sinharaja

Morapitiya Located on the way to Sinharaja, Morapitiya can at times be rewarding. Check the state of the access road, as this is highly variable.
Kithulgala Rainforest in the mid-hills, important as it provides an altitudinal gradient between the lowlands and highlands.

Montane Zone
Horton Plains National Park Important site for montane flora. Consider a visit to Hakgala Botanical Gardens.

Dry Lowlands (South)
Yala National Park Yala is best known for Leopards, and Sri Lankan Elephants and Sloth Bears. It is also a good site for dry-zone flora.
Uda Walawe National Park Grassland and monsoon forests. There are some fine examples of dry-zone trees standing in isolation, but there is no access on foot.

Dry Lowlands (North-central)
Mannar Island, Kalpitiya Peninsula Extensive stretches of mangroves.
Minneriya and **Kaudulla National Parks** For the Elephant Gathering, which peaks in August and September. There are also some fine stands of monsoon forests, and many fine specimens of trees such as ebony.

Mid-hills
Kandy Although Kandy in not on the itinerary for wildlife tours, the forests around it are botanically rich. In the town centre is Udawattakele, a forest park. Close by is the famous Peradeniya Botanical Gardens. A long day excursion is possible to the Knuckles Wilderness.

BOTANICAL GARDENS & ARBORETUMS

The Department of National Botanical Gardens website provides a listing of sites under its purview. The main botanical gardens are the Peradeniya Royal Botanical Gardens in the mid-hills in Kandy, the Hakgala Botanic Gardens in the highlands (close to Nuwara Eliya) and the Henarathgoda Botanic Gardens in Gampaha, in the wet lowlands over an hour's drive from Colombo. These botanical gardens are old and provide some fine examples of old trees. However, they have traditionally been somewhat ornamental in outlook and are biased towards having trees from other countries rather than showcasing native species. More recently created botanic gardens are the Dry Zone Botanic Gardens, Mirijjawila and the Wet Zone Botanic Gardens, Avissawella.

The Kottawa Forest Arboretum near Galle is one of the best places in which to learn to identify native species, especially the endemics of the wet zone, as it has a a footpath with labelled trees. The arboretum in Sinharaja is relatively new, being just a few decades old, and the labelled trees are small in height – but this does confer the advantage of the leaves being at head height or less, making them easy to inspect.

For dry-zone plants, consider a visit to the Popham Arboretum in Kandalama, near the archaeological and cultural site of Dambulla. The shared private grounds of the Cinnamon Lodge and Chaaya Village have some fine examples of labelled dry-zone trees. This is also the case with Sigiriya Village, a large tourist hotel near the archaeological site of Sigiriya. Although the gardens are not public, you should be able to visit them to use the restaurant and enjoy the trees. Many of the leading hotel chains in Sri Lanka, such as Aitken Spence, John Keells and Jetwing, have knowledgeable naturalists who can assist botanists.

Photographing Plants In the government botanic gardens such as Hakgala, which are used to having a lot of birders arriving with telescopes and tripods, the use of a tripod should not be an issue. Some of the other botanic gardens may charge an extra fee for using one (perhaps because it is assumed that anyone with a tripod is taking pictures for commercial purposes). There are generally no restrictions on using cameras. Colombo's Viharamahadevi Park, under the purview of the city municipality, has some fine examples of trees, although these are mainly foreign imports that also now adorn the city streets. The park does not permit photography, whether it be on the grounds of privacy for other visitors, security or restrictions on commercial use. In the national parks administered by the Department of Wildlife Conservation and the Forest Department, you are free to photograph the plants and animals you see, but even in Horton Plains, where you can go on foot, the trees are not labelled.

Sri Lankans are very friendly and outside the major cities, if you ask, people will be happy for you to even come into their gardens to photograph a plant if something catches your interest.

TAXONOMIC CLASSIFICATION

Identifying species in a group, whether it is birds, mammals, reptiles or butterflies, becomes easier when you become familiar with what 'type' of animal it is. Scientists group all species into taxonomic hierarchies, and a simple hierarchy would be one where related species are grouped into genera that are in turn grouped into a family and in turn into an order. The tree of life attempts to explain the inter-relationship of species from an evolutionary point of view.

The number of levels, or taxonomic ranks (like families and orders), used can vary for different groups of living things. Sometimes different authors will have different views on the levels in the hierarchy. The various ways in which scientists attempt to classify living things is the science of taxonomy and is outside the scope of this book. Controversy and debate rage on how best to do this, and as a result different books often classify species differently. Classification is also seemingly complicated by the use of suborders, subfamiles and superfamilies, which different authors utilize to arrange species into hierarchies. Animals such as butterflies become even more complicated with the use of tribes. However, these are all ways of trying to map species into a tree of life to understand how they are related to each other through the act of evolution over time. Most people can ignore the classification hierarchies and be content with simply getting a feel for where things approximately belong in terms of simpler relationships at family and species level.

In this book, all the plants that are covered fall into 11 clades. These are early angiosperms, magnoliids, monocots, commelinids, eudicots, pentapetalae, fabids, malvids, super asterids, lamiids and campanulids. Those who wish to understand more can find further information on the website for what is known as APG4. This is the fourth edition of the classification by the Angiosperm Phylogeny Group.

The common name of a species is followed by its Latin name, which generally has two parts, the genus and the specific epithet. No two species can have the same Latin binomial, and the Latin names are relatively stable, with a species having only one accepted Latin name at any given time. Common names, on the other hand, can vary widely from one country to another, or even within regions in a country. Some species have trinomials, where they have been described as subspecies or geographical races. Among mammals in Sri Lanka, for example, the Purple-faced Leaf Monkey *Trachypithecus vetulus*, an endemic species, has four distinct subspecies. It logically follows that all four subspecies are endemic. As another example, among dragonflies, the one found in Sri Lanka known as the Sri Lanka Cascader *Zygonix iris ceylonicus* is found only as a single subspecies, which is endemic. It has subspecies status to distinguish it from other populations found elsewhere in the world, similar enough to be considered the same species but different enough to be treated as distinct subspecies. Islands often have subspecies that are endemic as a result of isolation.

Endemic subspecies on islands are of great interest to 'listers' as they are candidates for a split into a 'good' or new species. Listers are wildlife enthusiasts like birders who want to maximize the number of species and especially the number of endemic species they have seen at home or abroad. Especially when travelling abroad, they will want to see an endemic subspecies in case it is split later into an endemic species.

Plants have many varieties, usually when they have been bred for horticulture. However, wild plants can also have naturally occurring varieties or subspecies. Recognizing subspecies in plants is more tricky than in animals unless there are clear geographical boundaries. This is because the influence of epigenetics, where the environment interacts with the instructions in DNA, is more obvious in plants. Seeds from a plant that takes root in open, sunny and poor soil can look very different from seeds that have taken root tens of metres away under shade in nutrient-rich soil – they can look like two different species. A plant may grow as a shrub on an exposed mountaintop, and as a tree in a more sheltered environment.

■ THE PARTS OF A FLOWER ■

CLASSIFICATION

The taxonomic hierarchy of the Common Helmet Orchid, for example, will look like this.

Division:	Angiosperms
Unranked clade:	Monocots
Order:	Asparagales
Family:	Orchidaceae
Genus:	*Polystachya*
Species	*Polystachya concreta*

What are known as seed-bearing plants have two natural divisions: gymnosperms (the conifers) and angiosperms (the flowering plants). Gymnosperm means naked seed and refers to the seeds or ovules not being enclosed in a carpel. Angiosperms (or covered seeds) are flowering plants in which the ovules are enclosed within an ovary. Traditionally, the flowering plants were split into monocotyledons and dicotyledons. Monocotyledons are plants with one seed leaf in the embryonic plant and include plants such as grasses, in which the leaves are thin, long and have parallel veins.

The dicotyledons, with two seed leaves in the embryonic plant, are also known as broad-leaved plants and are the rest of the seed plants. This division is, however, not a natural grouping, as it is not a result of a simple, two-branched evolutionary split. Phylogenetic analyses show that ancient dicotyledonous plants split into monocotyledons and into a further branch of new dicotyledons. Thus the flowering plants we see today are made up of a mix of old dicots (such as magnolias), monocots and newer dicots (such as sunflowers). In taxonomic terms the monocots are monophyletic, meaning that everything classified as a monocot has a single common ancestor from the branching events that led to the older dicots branching into monocots and more modern dicots.

THE PARTS OF A FLOWER

In a flower, the male part consists of the stamen, which has pollen-bearing anthers on top of a filament. The female part comprises a stigma that is a pollen receptor and a tube called the style, which leads into the ovary. This contains one or more ovules that when fertilized develop into seeds. In botanical terms, the fruit is the enlarged ovary containing the seed or seeds, but more generally the term fruit refers to the whole structure containing the seeds, which can include a fleshy or hard covering around the seed-bearing ovary.

The ovary, style and stigma are collectively known as a carpel. If a flower has only one carpel or has free carpels, these individual carpels are referred to as a pistil. Where a flower has two or more fused carpels (syncarpous), the term pistil refers to all of them. When pollen lands on the stigma, it grows a thin tube along the style to the ovary to fuse with the female cells (ovules), and starts to develop as a seed. Below the petals are small green parts – the sepals that cover the flower as a bud. In some flowers there is no clear distinction between the sepals and petals, and these are termed tepals.

The sepals and petals are known as the perianth. If the ovary is below the perianth it is known as inferior, and if it is above it is superior. In some families (like the capers), the

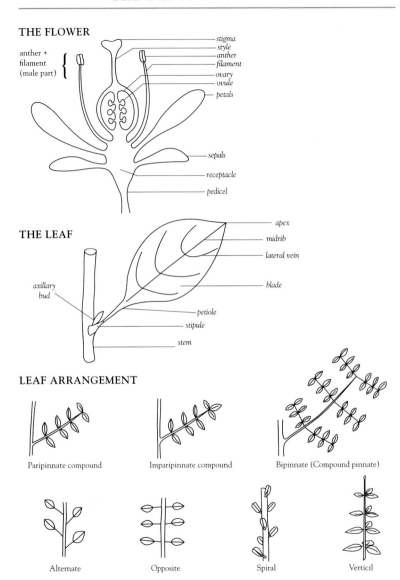

THE FLOWER

anther +
filament
(male part) {

- stigma
- style
- anther
- filament
- ovary
- ovule
- petals

- sepals
- receptacle
- pedicel

THE LEAF

- apex
- midrib
- lateral vein
- blade

axillary
bud

- petiole
- stipule
- stem

LEAF ARRANGEMENT

Paripinnate compound

Imparipinnate compound

Bipinnate (Compound pinnate)

Alternate

Opposite

Spiral

Verticil

INFLORESCENCE TYPES

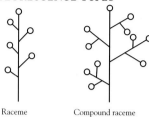

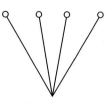

Raceme

Compound raceme
(panicle)

Spike

Fascicle (cluster)

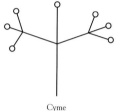

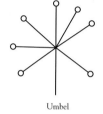

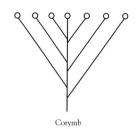

Cyme

Umbel

Corymb

FLOWER ARRANGEMENT

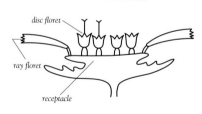

disc floret

ray floret

receptacle

Capitulum daisy style

Terminal flower

Axillary flowers

LEAF SHAPES

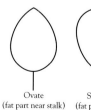

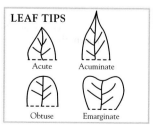

Ovate
(fat part near stalk)

Subovate
(fat part furthest
from stalk)

Lanceolate

LEAF TIPS

Acute

Acuminate

Obtuse

Emarginate

ovary is raised well above the perianth on a stalk known as a gynophore.

Dioecious plants are those with separate male and female plants – that is, the plant will only contain female flowers or male flowers. This makes cross-pollination easy as there is no risk of self-pollination. Monoecious plants have both male and female flowers, or bisexual flowers (containing both male and female parts). One strategy to avoid self-pollination is to have the male and female parts mature at different times. If both the male and female parts mature at the same time, plants wishing to avoid self-pollination have to develop other mechanisms. However, some plants opt for self-pollination. This can happen, for example, in environments where pollinators are absent.

The ovary contains the ovules, which after fertilization develop into an embryo or young plant; this is neatly folded within a seed. As already mentioned, flowering plants are known as angiosperms, which means 'enclosed seeds'; a reference to the seeds being enclosed in an ovary, which develops into a fruit. By contrast, gymnosperms ('naked seeds') do not have the seeds growing within an ovary. Fruits can take many forms. In popular parlance the term fruit is used in cases where there is an edible, fleshy covering. However, in botanical terms a nut is also a fruit. A fruit such as the pineapple is actually a compound fruit made up of many fruits. Thus the popular and botanical meanings of fruit can be different.

The edible figs produce what are popularly referred to as fruits. However, the fleshy part that we, and birds and other animals, eat is actually a fleshy receptacle upon which (on the inside of the fig) tiny flowers grow out of human sight. There is a tiny hole in the fig through which pollinating fig wasps enter. The fertilized ovules of flowers develop into seeds. Thus the normal cycle from ovules to seeds occurs hidden from view. In many plants the carpels split (or dehisce) into the component carpels. These are often contained within a fleshy covering, which is perceived as a fruit.

POLLINATION OF PLANTS & SEED DISPERSAL

For the next generation of plants to be produced through sexual reproduction, the ovules must be fertilized. As mentioned earlier, this requires pollen from the stamens to be deposited on the stigma. Plants achieve pollination using three main vectors – wind, animals and water. For pollination by wind, the pollen must be very light and produced in large quantities to achieve results. To use animals, the plants have to provide food, usually in the form of nectar, to offer animals an inducement to visit the flowers and to unwittingly carry the pollen to another flower. This has given rise to many complex and intricate relationships between plants and their pollinators through co-evolution. Pollination by water happens in two ways. Some aquatic plants produce light pollens that float through the water's surface and reach a stigma. Some other species, especially submerged ones, produce heavier pollens that sink and are caught by the stigma.

Plants also use the three elements of wind, animals and water to disperse their seeds so that they may find the right physical conditions in which to grow into new plants. Once again, where animals are used for dispersal, the plants have to offer a reward, and this is in the form of energy-rich food covering the seed or in the seeds themselves being edible.

GLOSSARY

Technical terms have been avoided as much as possible in this book. However, to avoid at times long-winded explanations, standard botanical terms have been used. These are included in this glossary, together with a number of other standard botanical terms that are useful to know when referring to more advanced texts.

achene Single-seeded fruit, dry and with a thin wall, not splitting when ripe.

actinomorphic Radially symmetrical. The flower can be divided into two mirror-image halves by dividing through the circumference.

acuminate Tapering to long tip (for example at leaf tips).

adaxial Refers to side of an organ that bears in the direction of the axis to which it is attached. Upper surface of a leaf is adaxial to stem, and undersurface is abaxial.

adventitious roots Roots from somewhere other than root system.

aerial roots Roots that arise from above the ground. In the case of some mangrove plants these are in addition to the main roots, which are underground. Epiphytic plants also have aerial roots, which dangle and are not connected to soil.

alternate Refers to leaves that are spaced apart on opposite sides and not directly opposite each other.

androecium Male parts of a flower comprising stamens. Each stamen consists of an anther and filament.

androgynophore Stalk carrying both stamens and ovary, arising from above where petals are joined, for example in passion flowers (*Passiflora* spp).

anisophylly Pair of opposite leaves at node with a pronounced difference in size or shape.

appressed Lying flat and close to stem, for example in the case of hairs or branches. Can also refer to being pressed against the ground for plants that are appressed to the soil, like the sundews.

auricle Ear-like projection at base of leaf, leaf blade or bract.

axil Angle between stem and leaf stalk.

axillary Arising on axil, as in the case of stipules or flowers, between angle of a leaf stalk and stem.

bifid Split in two, for example apex of a leaf, split or divided into two.

bisexual Having both male and female parts in the same flower.

bole Unbranched part of tree trunk (above buttress if one is present).

bract Small modified leaf at base of inflorescence or flower stalk.

bracteole Small modified leaf below flower or between bract and flower.

bulb Underground stem, as in onions.

caducous Tendency of leaves (including stipules) to fall off easily or prematurely. Not persistent.

calyx Outermost whorl of flower's organs, often divided into sepals.

capillary Slender, nearly hair-like.

capitulum Densely packed head of stalkless flowers arising at level of a flattened axis, as in daisies.

caudate Refers to leaves with pronounced drip tip. Leaf shape ends abruptly with extended thin tip.

chartaceous Thin and stiff; like paper.

ciliolate Fringed with very small hairs, for example the calyx of *Derris parviflora*.

circinate Coiled inwards, as with circinate tendrils.

clavate Club shaped.

connate Joined or attached to. Refers to similar parts, for example connate petals that may be fused at bases.

cordate Refers to leaf that is deeply notched at base to form heart shape.

coriaceous Refers to leaves that are stiff and leathery.

corm Underground swollen stem used for storage, for example that of *Colocasia* spp.

corolla Second whorl of a flower's organs, inside or above calyx, and outside stamens. Corolla is formed of either fused or separate petals.

corona This is formed by a series of appendages on petals of corolla, or on backs of stamens, or at junction of corolla tube and petals. The appendages are often united to form a ring, as in the Passifloraceae. See also Amazon Lily *Eucharis grandiflora*.

corymb Type of racemose inflorescence where flowers form flat top, with inner flowers (the higher ones) being on shorter stalks than outermost (lower) ones.

crenate A crenate leaf margin is one notched with regular, rounded, symmetrical teeth.

crenulate Margins with small, rounded symmetrical teeth.

culm Stem of a grass or sedge.

cyme Inflorescence where central flower is oldest. Later flowers arise from leaf axils in central flower, and other flowers arise from leaf axils in these. Inflorescence can be a simple or compound cyme. A thyrse is where a raceme has flowers replaced by cymes.

deciduous Refers to plant that seasonally sheds leaves. In high latitudes plants shed leaves in winter, and in the tropics they do so during dry periods.

decurrent Term applied when leaf blade runs down leaf stalk and it is difficult to separate the two. Also used when leaf blade forms wing around petiole or stalk.

decussate Alternate opposite pairs of leaves are at right angles to each other.

dehiscent Splitting or dehiscing along predetermined lines when mature, applied to fruits and anthers.

deltoid, deltate Shaped like equal-sided triangle. The term deltate is preferred. Mostly applied to shape of leaf.

dentate Toothed margin (typically of a leaf) with prominent, symmetrical, sharp points.

denticulate Finely toothed. A denticle is a fine tooth.

depressed Flattened from above downwards, as in fruit of Cherry Guava *Psidium cattleyanum*.

dichasial cyme Where an inflorescence has a central stalk with a terminal flower, laterally and opposite to each other are two branches with a terminal flower, and again each of these has a pair of lateral and opposite branches bearing flowers.

dicotyledon Term applied to plants with two seed leaves in the embryo.

dioecious Plants with unisexual flowers, with the male and female flowers being on separate plants. Thus each plant will be either a female or male plant.

disc floret In the Asteraceae (daisy family), a flower head comprises disc florets in the central disc, and on the rim are ray florets that look like petals.

distal Refers to further end from point of attachment. Distal end of a leaf will be the leaf tip (as opposed to leaf base).

distichous Arranged in two vertical rows.

domatia Structures, usually in the leaf, but also in the stem or root, inhabited by animals, especially ants.

drupe Fleshy, indehiscent (not splitting) fruit with seed having hard or stony covering (endocarp), surrounded by fleshy pulp, as in a cherry.

emarginate Notched at tip, for example of a leaf.

ensiform Refers to long, narrow leaves, ending with a point, for example pineapple leaves.

entire In the context of leaves, refers to an undivided margin. Leaf margin is smooth without crenulations, lobes and so on.

epicalyx Whorl of bracts beneath flower that looks like a second calyx. Common feature in shoe flowers.

epipetalous Usually refers to stamens that are united with petals and appear to lie on petals.

ericoid With needle-like leaves, typical of heathland plants.

exserted Projecting, as from a sheath or pod.

exstipulate Without stipules.

fascicle Cluster of similar organs all arising from same point on plant, for example a cluster of leaves, fruits or stamens.

flower Reproductive organ of a plant. Part of a plant that has male or female parts, or both, capable of reproduction. What looks like a flower to a non-expert may to a botanist's eye be a collection of flowers and be more correctly an inflorescence.

foliolate With leaflets.

follicle Dry fruit formed from single carpel that splits along one side (axis) only.

fruit Strictly speaking, the fertilized mature ovary of a seed-bearing plant, but more widely used to include berries, drupes and compound fruits

fusiform Spindle shaped, thick in middle and tapering at both ends.

gamopetalous With fused petals, at least at base.

gamosepalous With fused sepals, at least at base.

glabrous Smooth, that is without hairs or other epidermal growths. A glabrous leaf surface is smooth and shiny without hairs.

gregarious Refers to plants occurring in communities, such as dipterocarp communities in lowland rainforests.

gynodioecious Refers to plants bearing either female or bisexual flowers.

gynoecium Female parts of a flower comprising one or more fused or free carpels. Each carpel contains an ovary, style and stigma.

gynophore In some families (like the capers), the ovary is raised well above the perianth on a stalk known as a gynophore.

halophytes Plants adapted to salty conditions.

hastate Where bases of a leaf comprise two approximately triangular projections or lobes that are outwards pointing. Hastate leaves can be arrow shaped.

herbaceous Herb-like, that is a non-woody plant.

hermaphrodite Bisexual flowers where male and female parts are found in same flower.

heterostyly Plants that have flowers with two or more different lengths of style – an adaptation to improve pollination success.

hirsute With coarse and rather stiff hairs.

hispid With long, stiff hairs or bristles. More bristly than hirsute.

imparipinnate Pinnate with single terminal leaflet. Thus, there is an odd number of leaflets.

indehiscent Refers to fruit that does not split naturally on ripening.

indigenous Occurring naturally in a country or region without having been introduced by people.

indumentum Covering of hair or scales.

inferior ovary Sepals and petals are known as the perianth. If the ovary is below the perianth it is known as inferior, and if above as superior.

inflorescence Collection of flowers, where a flower is the part of a plant that has male or female parts, or both, capable of reproduction.

infundibuliform Funnel shaped; applies especially to styles.

involucre Whorl of bracts beneath an inflorescence.

keel petal In pea (Fabaceae) family, one large, showy petal is turned upwards and is the keel. Other petals are fused to form an inverted boat-shaped keel (standard) that encloses the stamens and style to form a two-lipped flower.

labellum Lowest petal in an orchid, different from other two lateral petals, and often larger.

lancelolate Ovate or egg shaped at base and narrowing at upper half to a point.

lenticels Spots on bark, which may be circular or oval shaped; may be loosely arranged into rings around bark. Cherry trees grown in parks have pronounced lenticels.

loculicidal As in a seed-containing capsule that splits longitudinally along dorsal sutures of the wall.

-merous Refers to main parts of a flower (sepals and petals) appearing in groups of three, four, five and so on (3-merous, 4-merous, 5-merous and so on). In some flowers, sepals, for example, may be trimerous and petals pentamerous.

moniliform Refers to root that looks like a string of beads.

monocotyledons Plants such as grasses with one seed leaf in embryo. In adult plant leaves have parallel veins and flowers are trimerous (in sets of three).

monoecious Having male and female flowers or bisexual flowers in one plant.

monotypic Containing only one species in a genus or only one genus in a family.

mucilage Slimy excretion that swells on absorbing water.

mucronate Refers to leaf that ends with short, stiff point or tip.

naked flowers Flowers without a calyx or corolla.

ob- As a prefix, has two meanings: 'against', and in botany used to indicate that an attachment point is opposite to the usual shape.

oblanceolate Refers to narrow, parallel-sided leaf, rounded at top and base.

opposite Refers typically to a leaf arrangement where leaves are directly opposite each other on a stem, or in a compound leaf where leaflets are directly opposite each other on an axis. (Alternate leaves are spaced apart on opposite sides and not directly opposite to each other.)

orbicular Flat and circular in shape (a sphere will be described as globose).

pachycaul Thick stemmed and sparsely branched.

panicle Can be thought of as a compound raceme. An inflorescence where main stalk has other branches (which can in turn be branched) carrying flowers in a raceme, as in flowers of grasses.

papillae Soft, small protuberances that extend beyond or above the surface.

pappus Series of bristles or hairs found on base of corolla that are later found at tip of fruit, as in the Asteraceae family.

paripinnate Pinnate leaflets without a terminal leaflet; an even number of leaflets in a compound leaf.

patana, patna Grassy hillside.

pedicel Flower stalk of an individual flower in an inflorescence.

pedicellate Stalked flowers.

peduncle Has multiple meanings. In an inflorescence, lower unbranched part of stalk is called a peduncle, and upper section with branches of flowers is a rachis. Also used to refer to stalk of single flower, or common stalk of flowers without stalks (that is, sessile flowers).

peltate Round and circular, and attached in centre, as in leaves of a water lily, where petiole is not attached to blade margin but in centre.

pendulous Hanging, like a pendulum.

perianth Floral envelope comprising outer whorl of calyx (made of sepals) and inner whorl of corolla (made up of petals). One or both whorls may fuse to form a tube.

pericarp Term that includes epicarp (skin), mesocarp (middle) and endocarp (inside) layers of a fleshy covering around a seed, for example in a single-seeded fruit like a mango.

persistent In the case of stipules, used to denote that they do not fall off easily after new leaves have grown. Generally refers to parts of a plant that do not fall off easily as expected.

petaloid Like a petal, for example in the case of sepals that are coloured and shaped like a petal.

phloem Plant tissue used for transporting food material. In a woody bark, phloem will be inside. Note that the xylem is the tissue that transports fluid.

pilose Hairy with short, thin hairs.

pinnae Leaflets on either side of the rachis (or midrib) of a compound leaf. Usually refers to first order division of a pinnate leaf, which may have further divisions.

pistil Where carpels are free (that is, apocarpous), or fruit has a single carpel, refers to female reproductive part of a flower comprising stigma, style and ovary. In this situation, carpel and pistil refer to the same thing. Where carpels are fused (that is, syncarpous), the term pistil refers to all female organs in a flower. Another term for this is gynoecium.

pollinia Club-shaped masses of sticky pollen found in orchids.

polygamous Having male, female and bisexual flowers on the same plant.

prostrate Term used for plants lying flat along the ground.

protandrous Stamens (male parts) in bisexual flower mature before female parts are receptive (this avoids self-pollination).

protogynous Stigmas (female receptors) in bisexual flower mature before male parts are ready to disperse pollen (this avoids self-pollination).

psammophytes Plants adapted to growing in sand or sandy soils.

pseudobulb 'Above-ground' storage organ. Found in many species of epiphytic and sympodial orchid.

pubescent Hairy. A pubescent leaf is one covered with downy hairs.

pulvinate Swelling on petioles at join with stem, or at join with leaf blade, or both. Dipterocarps, for example, have pulvinate petioles.

pulvinus Swollen or enlarged section of leaf stalk that allows plants to change direction (sleep movements) of their leaves in response to light.

pyriform Pear shaped.

raceme Term used where flowers are attached by pedicels to a single stalk. Similar to a flowering spike, but flowers are on individual stalks that come off a central stalk. There are many variations to inflorescences arranged in a raceme. In a spadix and a spike, flowers are sessile (without a stalk). In a raceme, newest flowers are at top and are last to open.

rachis In a compound leaf, part of central axis, between first set of leaflets and tip, excluding part of axis up to first set of leaflets, which is the petiole. Midrib or main nerve of a compound leaf. In an inflorescence, rachis is similarly part of axis, except the first part (the peduncle) up to the first set of flowers.

ray floret In the Asteraceae (daisy family), a flower head comprises disc florets in the central disc, and on the rim are ray florets that look like petals.

resupinate Refers to flowers that are upside down (like orchids), or seemingly so.

reticulate Refers to lace-like network of veins on a leaf.

rhizome Underground stem. Can be told apart from roots by presence of nodes, buds or scale-like leaves.

rotund Nearly round (or circular). In the context of leaves, assumes a nearly circular, two-dimensional shape.

samara Dry, non-splitting (indehiscent) fruit with a wing that is longer than part with seed.

samaroid Resembling a samara, but with wing surrounding seed chamber.

saprophytes Plants that lack chlorophyll and cannot manufacture their own food. They need to absorb nutrients from decaying organic matter. Some orchids are saprophytic.

scales Modified leaves that protect other parts in a plant, such as the growing point, from frost. Less common in warm tropics than in cooler areas.

scape Leafless flower stalk or inflorescence stalk.

scarious Having a dried-up appearance.

sepaloid When petals look like sepals (and are functioning as sepals).

septicidal Septicidal dehiscence is where ripe (mature) capsule splits along lines of carpel junctions, but the valves (parts of carpel that open out) do not fall off and remain attached. In septifragal dehiscence valves fall off and an axis (the columella) to which seeds are attached remains.

serrate Saw toothed, with regular, acute-angled teeth (pointing towards apex).

serrulate As in serrate but with minute teeth.

sessile For example, leaves without a petiole would be sessile leaves.

shrub Woody plant that is much branched near base (a tree, on the other hand, grows up a single straight trunk to some height before branching).

simple In the context of leaves, simple leaves are those not divided into leaflets.

spadix Unbranched inflorescence with thick, fleshy axis. Flowers are attached to it and may in some cases be partially submerged.

spatulate Spoon shaped; also spelt spathulate.

staminode Sterile stamen, usually smaller than fertile stamen and not bearing pollen.

standard petal In pea family, one large, showy petal is turned upwards and is the keel. The other petals are fused to form an inverted boat-shaped keel (standard) that encloses the stamens and style to form a two-lipped flower.

stipel Stipule-like outgrowth occurring at base of a leaflet or pair of leaflets in compound leaves.

stipitate Supported on a special stalk; that is, not supported on peduncle, pedicel or petiole.

stipule Growth at base of leaf on leaf petiole, usually in pairs, which can be leaf-like or spine-like. Note that 'exstipulate' means without stipules.

stolon Vegetative shoot that spreads along the ground, giving rise to new plantlets at nodes.

sub Used as a prefix, with two meanings. The first refers to almost or nearly – for example, subdeltate is nearly deltate (triangular). The second meaning is 'below' or 'under'.

subulate Awl shaped, like stout needle tapering to fine point.

superior ovary Sepals and petals are known as the perianth. If the ovary is below the perianth, it is known as inferior and if it is above it, as superior.

sympodial Without a single main stem. Sympodial orchids, for example, do not have single main stems.

talawa Grassland with scattering of trees.

tank Man-made lake or irrigation reservoir.

terete Circular in cross-section.

thyrse Inflorescence where a raceme has flowers replaced by cymes.

tomentose With dense covering of short, soft hairs.

tomentum Covering of downy hairs, like felt.

trichome Hair-like outgrowth.

trifid Split in three.

trifoliate With three leaflets.

tristichous With leaves arranged vertically in three rows, one above the other.

tuber Underground modified stem or root, used for storage.

tubercular, tuberculate Covered with wart-like protuberances. Knobbly.

turbinate Shaped like a spinning top, with the top conical.

umbel Where stalks of individual flowers are joined together at their bases.

undulate Wavy, usually referring to leaf margin.

unisexual In reference to flowers, having only male or female parts.

urceolate Urn shaped – constricted at top and expanded again slightly to form narrow rim.

valvular dehiscence When fruit splits open along sutures of carpels.

velutinous Velvety; soft to the touch.

vermiform Worm shaped, thick and bent in places; especially of roots.

verticils Arrangement in a whorl of structures that is not typically found in whorls, as in flowers of Ceylon Slitwort *Leucas zeylanica*, arranged in whorls or verticils.

villous, villose With long, soft hairs.

viscid Sticky.

viscidum Glands to which pollinia in orchids are attached.

viviparous Bearing live young, a condition common in mammals and less rare in other animals. In plants, refers to seeds germinating on parent plants, as in the Rhizophoraceae.

whorl Refers to branches or leaves arising from same point on stem, like spokes on a wheel, and where there are more than two leaves together at the same level.

xylem Woody tissue used for transporting fluids in vascular plants (note that the phloem is the tissue that transports food).

zygomorphic Refers to bilateral symmetry; only a single vertical dividing line can be drawn so that the flower is divided into two vertical halves, each of which is the mirror image of the other. Term used for flowers that are vertically symmetrical, like orchids.

NYMPHAECEAE (WATER LILIES)
This distinctive family of freshwater plants is characterized by large, showy flowers on long stalks and large leaves. The plants appear to float on the water but are in fact rooted at the bottom, hence they occupy shallow ponds and lakes. Water lilies are found in tropical and temperate regions. They are one of the most primitive groups of flowering plants, with 60 species in six genera. The largest plants are in the genus *Victoria*, found in South America.

Blue Lotus ▪ *Nymphaea nouchali*
(S: Manel)

IDENTIFICATION Growth habit that of an aquatic herb. Leaves flat and circular (orbicular) or suborbicular, 30cm across, smooth and coriaceous. They are attached to the petiole in the middle (peltate). Flowers large, 10cm across, and fragrant. Sepals longer than petals, brownish-green underneath and pale violet on surface. Many pale violet or light blue petals. Numerous anthers; outer anthers pale purple, inner anthers yellow. The national flower of Sri Lanka. **DISTRIBUTION** Native. Common in dry lowlands. **HABITAT** Ponds and lakes. Also grown widely as a cultivated plant. **FLOWERING PERIOD** Throughout the year. Flowers open with sunrise and keep opening until early afternoon.

Hairy Water Lily ■ *Nymphaea pubescens*
(S: Olu)

IDENTIFICATION Aquatic herb with large, floating, flat, circular (orbicular) leaves with toothed margins. Leaf hairy underneath. Flower stalk (pedicel) slightly prickly. Flowers 15cm across, emerging slightly from the water. Sepals light green underneath, with surface similar to colour of petals. Many petals; white, purplish-pink, red or yellow. Outer petals larger than inner ones. Many stamens. **DISTRIBUTION** Native. Dry lowlands. **HABITAT** Ponds, lakes and reservoirs. **FLOWERING PERIOD** Throughout the year.

> ### MAGNOLIACEAE (MAGNOLIAS)
> The magnolias are a primitive family of angiosperms with a long phylogenetic branch length. They grow as deciduous or evergreen trees, or shrubs, and have simple or lobed leaves. They have a curious disjunct distribution between the Americas and Asia. Around three-quarters of the species are found between temperate East Asia and tropical Southeast Asia, while the genus *Taluma* is native only to the New World.

Nilgiri Chempaka ▪ *Michelia nilagarica* var. *walkeri*
(S: Wal sapu)

IDENTIFICATION Tree that grows to about 12m tall. Stem of young branches black. Stipules prominent and covered by silky hairs. Leaves elliptic, deep green and leathery, with acute tips. Flowers solitary, cream in colour. No distinct petals or sepals. Flowers sweet scented. **DISTRIBUTION** Endemic. Mountains at 1,200–2,350m. **HABITAT** Upper montane forests. **FLOWERING PERIOD** February, November.

ANNONACEAE (SWEETSOPS & SOURSOPS)
This family is found mainly in the tropics and subtropics, with a few species occurring in the temperate zone. It primarily comprises tropical trees, but some species grow as shrubs and lianas. There are about 2,500 species in around 135 genera, with the highest number of species occurring in the Old World, followed by the New World and Africa. The plants are commercially important for their edible fruits, including the cherimoya, sweetsop and soursop. Their simple leaves are usually alternate and borne in two ranks. The radially symmetric flowers are aromatic and often pendulous.

Bullock's Heart ■ *Annona reticulata*
(S: Anoda, Atta, Attha, Rata, Sitha, Weli-attha)

IDENTIFICATION Tree with erect and spreading trunk, about 2–3m tall. Leaves oblong in shape and alternately join with stem. Young stems and underneath of leaves crowded by dense hairs. Veins clear on both leaf surfaces. Crushed leaves smell bad. Flowers borne in a cluster, and each flower has three outer fleshy petals (fleshy petals are not common in

plants). Petals about 4–5cm long. Outer margins of petals brown and hairy, and inner area of petals whitish. Flowers never fully open. Fruit is a large (10–15cm diameter) compound type, with a flattened (depressed) base. Surface of young fruit green, turning yellowish-brown during ripening. Fruit is edible, with sweet-tasting flesh.
DISTRIBUTION Exotic. Home gardens and man-made habitats of Central Highlands. **HABITAT** Marshy or wet places with mild temperature, along riverbanks or lake edges. **FLOWERING PERIOD** March, April, May.

Aristolochiaceae (Birthworts)

Birthworts comprise a family of about 500 species that are found mainly in the tropics and warm temperate regions. They occur mostly as herbs or shrubs, though some species are climbing lianas. They have heart-shaped leaves and showy flowers, and the flowers often have a pungent smell, like rotting flesh, which attracts flies for pollination.

Dutchman's Pipe ▪ *Aristolochia ringens*

IDENTIFICATION Herbaceous vine that twines and climbs over other plants and herbs. Leaves have very long stalks, are almost kidney shaped, and are alternately present on climbing stem. Veins radiate towards blade from leaf base. Flowers are very unusual – not like typical flowers of other plants. They are 15–20cm long, and green in colour with characteristic purple markings. Each flower has a sack at the base or close to the floral stalk, the interior of which is woolly. Sack is like a floral tube and is divided into two at the end, forming two long lips, one very long and the other short. Stalk of flower four times longer than stalk of leaf. **DISTRIBUTION** Exotic. Lower to middle elevations of lowland rainforests, and lower elevations of wet mountain forests. **HABITAT** Wet places with moderate sunlight. Climbs onto vegetation at lower levels, rarely climbing to upper reaches. **FLOWERING PERIOD** Latter part of the year.

APONOGETONACEAE (CAPE PONDWEEDS)
All species in this group of perennial aquatic plants have a rhizome and are in a single genus. They are found in the Old World and northern Australia, and can survive extended dry periods as dormant tubers, coming to life after the onset of rains. The flowers have conspicuous tepals (that is, the sepals and petals are not clearly differentiated). The leaves are spirally arranged at the base, and the shapes differ between submerged and floating leaves. Their typically strong midvein and parallel longitudinal veins are intersected by cross-veins, creating a distinctive mesh pattern.

Floating Lace Plant ▪ *Aponogeton natans*
(S: Kekatiya)

IDENTIFICATION Aquatic bulb plant. Dark green leaves emerge from bulb. Leaf margins characteristically undulate. Plant is submerged, with inflorescence arising from bulb and growing above the water level. Uppermost end of inflorescence downcurved towards the water. Tiny flowers borne on upper areas of inflorescence. Inflorescence axis is white. Flowers light violet in colour. **DISTRIBUTION** Native. Waterbodies of low-country dry zone. **HABITAT** Clean, shallow waterbodies with moderately high temperature. **FLOWERING PERIOD** Throughout the year.

> ## Pandanaceae (Screwpines)
> These shrubs, climbers and trees are found in the tropics and subtropics, growing from sea level to altitudes of 3,000m. There are about 900 species in three genera, and some species in the genera *Pandanus* and *Frecynetia* are used locally as food. The plants grow in thickets or as single individuals. Their stems are ringed with leaf scars, and some species have adventitious roots that help prop them up. The leaves are arranged vertically in three (tristichous) or four rows, one above the other, and are armed with teeth on the margins and sometimes on the leaf blade – they can be formidable. The unisexual flowers are naked.

Dense-leaved Freycinetia ▪ *Freycinetia pycnophylla*
(S: Kolla)

IDENTIFICATION Huge, woody climber that runs along stems of large trees by using adventitious roots. Stem is leafy, and leaves are linear, about 30cm long and sheathed over stem. Prickles along leaf margin create a dentate margin that can badly cut the unwary. The picture shows the ripe berries, which are scarlet/red and attractive. Flowers are tiny, and attached to club-shaped spadices, of which there are usually three per inflorescence. Separate male and female flowers occur in this species. Spadices are surrounded by orange-red bracts. **DISTRIBUTION** Endemic. Wet-zone and hill-country forests. **HABITAT** Very common on trees of wet forests. **FLOWERING PERIOD** March–April.

> **COLCHICACEAE (COLCHICUMS)**
> These perennial herbs are found in warm, Mediterranean-type climates, and there
> are about 225 species in 18 genera across the world. Typically, they have underground
> rhizomes or corms, although a few species have tuberous roots. They have
> characteristic large flowers, and long, narrow leaves. The floral arrangement can vary
> from racemes to cymes, and the flowers are radially symmetrical and bisexual, with six
> tepals and six stamens. Some species have stamens with nectar glands at the bases.
> The family includes a number of plants popular in gardens, such as the Flame Lily
> *Gloriosa superba*. Many species store highly toxic alkaloids.

Canton Fairy Bells ▪ *Disporum cantoniense*

IDENTIFICATION Herbaceous perennial that grows to about 1m tall. Simple leaves large
(nearly 6–7cm long), with three or five clear parallel veins on surface. Solitary flowers open

towards the ground, and are showy, white and
about 3cm in diameter. Each has six petals and six
anthers. Petal tips pointed (acute) and greenish
in colour. Anthers also green. Fruit of plant is
very attractive and colourful. Fruits blackish-blue,
with a shiny surface, and about 1cm in diameter.
DISTRIBUTION Native. Can be seen in higher
elevation areas of 1,100–2,900m, like Horton
Plains National Park and other mountain forest
areas. **HABITAT** Cool, shady places at high
elevations. **FLOWERING PERIOD** April–July.

Glory Lily ▪ *Gloriosa superba*
(S: Niyangala; T: Kartikaikilanku)

IDENTIFICATION Herbaceous climber. Leaf arrangement on stem is spiral. Leaves lack
stalks (sessile) and are very long. Leaf tips extended by tendrils. Flowers very attractive

and large, and open towards the ground. Petals or perianth parts
upcurved. Five sepals and petals in a flower. Base of perianth
(petals and sepals) yellow, with red tips. Below perianth there
are five stamens with long filaments and orange anthers. Ovary
green and clearly visible at bottom of flower. Fruit is a capsule.
Plant has an underground rhizome as a true stem. Plant contains
lethal toxins. **DISTRIBUTION** Native. Wet-zone forests and
moderately dry forests in lowlands. Also lower areas of montane
forests. **HABITAT** Common in forest edges and on waste ground.
FLOWERING PERIOD September–January. In wet zone, flowers
throughout the year.

ORCHIDACEAE (ORCHIDS)

This is the most species rich family of flowering plants, with 15,000–20,000 species in more than 1,000 genera. Orchids have distinctive zygomorphic flowers, with the flowers of most species having a pronounced lip and hood. They are also famous for their mycorrhizal fungi, which are essential for their growth. The seeds are tiny and like dust. Orchids have three basic growth forms: as terrestrial plants, epiphytes and saprophytes (plants that are parasitic and lack chlorophyll). Despite the huge profusion of species in the family, they are all perennial herbs. Their leaves are long, simple and alternate. Orchids have a cosmopolitan distribution. The epiphytic orchids are generally found in forests, in contrast to the terrestrial orchids, which are often found in open flowering meadows.

UNDERSTANDING ORCHIDS

The growth form of orchids can be monopodial or sympodial. In the former they have one main stem, and in the latter there are two or more stems. In monopodial orchids the same stem continues to grow season after season, and the flowers appear from the leaf axils. The monopodial orchids have no organs for the storage of food, and are tropical in distribution (like *Vanda* species). The sympodial orchids have pseudobulbs, which are swollen stems used for storing food. In seasonally dry climates they can swell considerably. Despite the incredible number of species, all orchid flowers are built around the same body plan of three sepals and three petals. Furthermore, the stamens are fused with the style and stigma to form a 'column'. The topmost sepal, the 'dorsal' sepal and the two lateral sepals protect the flower bud before it opens, and are often green until the flower opens. The two lateral petals are similar, with the third positioned opposite the dorsal petal and known as the labellum, or lip. The lip can be divided into two or even three lobes.

All orchids are bisexual and have a column where the male and female parts are fused together. The column may stand upright and be protected by the dorsal sepal and lateral petals, or lie parallel to the lip and be encased by the side lobes on the lip. On top of the column in the anther cap with the pollinia are sticky bodies with pollen, and below it lies the stigma, which is a sticky patch that receives pollen from other flowers. In most orchids the flowers are rotated through 180 degrees, so they are positioned upside down with the lip at the bottom.

The genus *Eria* resembles *Dendrobium*, but a key difference between them is that *Eria* has eight pollen masses and *Dendrobium* has four.

Aerangis hologlottis

IDENTIFICATION Stemless epiphytic orchid. Pseudobulbs absent. Leaves about 9cm long, oblong to oblanceolate, coriaceous and dark green with a shiny surface. Midrib prominent and slightly grooved on upper surface. Inflorescence is a 13–16cm-long raceme. White flowers about 1.5cm across. Dark brown bract at base of each peduncle. Sepals and petals three veined, and petals slightly larger than sepals. Lip lanceolate to spatulate in shape, and five veined. Cylindrical spur of lip almost the same in length as peduncle. **DISTRIBUTION** Endemic. Submontane and mid-country evergreen forests to about 500m. **HABITAT** Rare species found on trees of submontane and mid-country evergreen forests. **FLOWERING PERIOD** January–April.

Wanaraja ▪ *Anoectochilus setaceus*
(S: Wanaraja)

IDENTIFICATION Grows as a small, terrestrial, leafy herb with a creeping stem. Terminal end of stem erect and devoid of leaves. Each stem has 3–5 leaves, which are spreading, and

ovate in shape with an acute tip. Leaf colour conspicuous, helping to identify this species. Leaf surface dark velvety-green above with orange reticulations; pale underneath. Leaf petiole forms sheath on stem. White flowers borne on a terminal spike. **DISTRIBUTION** Endemic. Tropical wet evergreen forests extending to subtropical montane forests at 300–2,000m. **HABITAT** Common under shade of trees among fallen leaves. **FLOWERING PERIOD** January, May–September, December.

Bamboo Orchid ■ *Arundina graminifolia*

IDENTIFICATION Land orchid that grows to nearly 2m tall. Stem is a 'pseudo stem' and looks like a bamboo tree. Leaves look like grass leaves, and are alternately positioned along stem. Flower stems emerge from top of pseudo stem, which produces several flowers sequentially – thus one or two flowers are in bloom at once. Flowers white mixed with light pink-purple. Floral lip purple-pink with yellow patch at middle. Widest point of flower is 5–6cm. **DISTRIBUTION** Exotic. Nawalapitiya, Galaboda, Gampola, Kandy, Sinharaja Forest, Ratnapura and many other locations in mid-hills of wet zone. **HABITAT** Commonly seen in open wetlands in wet-zone locations. **FLOWERING PERIOD** Throughout the year.

Thwaites's Bulbophyllum ■ *Bulbophyllum thwaitesii*

IDENTIFICATION Epiphytic orchid with pseudobulbous stem. Internodes in between pseudobulbs. Pseudobulbs measure 0.8–1.3 x 0.5–0.6cm, and are subglobosely ovoid, tapering to apex, and ridged lengthwise. One leaf per pseudobulb. Petiole very short. Blade (about 6cm long) oblong, notched at tip (emarginate), coriaceous and stiff. Inflorescence is an umbel arising from base of a pseudobulb. Small flowers pale greenish-yellow. Peduncle green with red streaks. Dorsal sepal small (4mm long), broadly ovate and acute. Lateral sepals (11mm long), about three times longer than dorsal sepal. Lateral sepal tips obtuse. Petals smaller, with length similar to that of dorsal sepal. Lip about 2mm long and fleshy. **DISTRIBUTION** Endemic. Tropical wet evergreen forests extending to subtropical montane forests to 2,000m. **HABITAT** Rare species found on trees in tropical evergreen forests. **FLOWERING PERIOD** January, March, August, November.

The Fragrant Coelogyne ■ *Coelogyne odoratissima*

IDENTIFICATION Pseudobulbous epiphyte. Pseudobulbs subglobose, green, 1.7cm long x 1.4cm in diameter, and formed near flowering shoot. Two leaves per pseudobulb. Leaves are

smaller when flowers are present, and become larger when there are fruits in the flowering shoot. Maximum length of mature leaves about 13cm. Leaves green, linear, acute and coriaceous. Flowering stalk arises from base of a pseudobulb. Basal sheaths at base of flowering shoot. Each flower stalk has 2–4 fragrant white flowers, 3.5cm across. Petals smaller than sepals. Lip mixed with yellow and brown patch. Fruit is a capsule. **DISTRIBUTION** Native. Montane forests above 2,000m. **HABITAT** Common on trees in mountain forests. **FLOWERING PERIOD** December, January, March.

Ceylon Coelogyne ■ *Coelogyne zeylanica*

IDENTIFICATION Epiphytic orchid. Rootstock creeping and roots slender. Pseudobulbs small (1–1.2cm long), green, wrinkled, with 1–2 leaves. They become reddish-brown when

old. Leaves (7–12cm long) deep green, linear-lanceolate, coriaceous and acute. Flowers white with yellow patch in middle of lip. Flowers open towards the ground. One or two flowers grow on each inflorescence and flower stalks arise from bases of old pseudobulbs. Brown sheaths conspicuous in floral stalk, which is short (2.5–3.7cm) in comparison to floral stalks of other *Coelogyne* species. Fruit is a fusiform or spindle-shaped capsule, 1.7 x 1cm, and six ribbed.

DISTRIBUTION Endemic. Wet, moist cloud forests. **HABITAT** Very rare species that grows on trees in mist-laden cloud forests. **FLOWERING PERIOD** March, October–November.

Aloe Leaf Cymbidium ■ *Cymbidium aloifolium*

IDENTIFICATION Stems of this epiphyte are crowded, giving a tussock-like appearance. Stems fleshy and covered by leaf bases that sheath stem. Needle-like, thin aerial roots. Leaves long (about 45cm), coriaceous, deep green, with tip unequally lobed. Raceme pendulous, about 30cm long. Flowers cream, stained reddish-purple down middle of sepals, and petals have purple-blotched lip. Sepals about 2cm long; petals about half as long as dorsal sepals. Fruit is a pyriform capsule about 5–6cm long. **DISTRIBUTION** Native. Mid-country to 1,220m. **HABITAT** Very common on tree trunks of tropical wet evergreen forests. Sometimes grown in home gardens. **FLOWERING PERIOD** March, April.

White Dove Dendrobium ■ *Dendrobium crumenatum*
(S: Sudu-pareyi-mal)

IDENTIFICATION Epiphyte with leaves arising at base like a tussock, with pseudobulbous stem and fibrous roots. Pseudobulbs about 7cm long and located a node or two above stem base. Leaves about 5cm long, coriaceaous and unequally notched at apex. Leaf base creates sheath around stem. Midrib very clear and prominent on underneath of blade. Flowers borne at terminal end of naked stem. Flower 4cm across, fragrant, and white with yellow patch in centre of lip. Lip broader, with wavy margin. **DISTRIBUTION** Exotic and cultivated in low-country wet zone. **HABITAT** Common and found mainly as cultivated form in wet zone. **FLOWERING PERIOD** March, December.

Fringed Dendrobium ■ *Dendrobium macrostachyum*

IDENTIFICATION Pseudobulbous stem thin, pendulous and slender, growing to 70cm long. Leaves sessile. Flowers greenish-yellow, 2–2.3cm across. A few flowers arise from nodes of leafless stems. Species can easily be identified by pink venation in lip. Fruit is a capsule 3cm

long. **DISTRIBUTION** Native. Low-country wet and dry zones, and submontane locations to 1,200m. **HABITAT** Tree trunks of wet-zone low country and dry-zone jungles. Common on trunks of coffee trees in home gardens, except in open areas. **FLOWERING PERIOD** March, April, June, July.

Lily of the Valley Orchid ■ *Eria bicolor*

IDENTIFICATION Grows as an epiphyte, with leaves arising at base like a tussock, and erect pseudobulbous stems. Roots clustered at bases of pseudobulbs, each of which has

nodes and is covered by brown, papery sheaths. Leaves clustered at end of each stem, and racemes arise from axils of old leaves. White flowers 2–3cm across. Fruit is an oblong cylindrical capsule. **DISTRIBUTION** Native. Submontane and mid-country tropical wet evergreen forests to 1,000m. **HABITAT** Very common on trees of wet and intermediate forests. **FLOWERING PERIOD** March, September–October.

Large Bracted Eria ■ *Eria braccata*

IDENTIFICATION Dwarf epiphytic orchid with globose pseudobulbous stem. Can be easily identified by presence of pair of leaves on tip of pseudobulb when flowering. Flowers 2.7cm

across and white (creamy-white). Lip and column have patches of yellow. Fruit is a 1.2cm-long, obovate-ovoid capsule. **DISTRIBUTION** Native. Submontane or mid-country tropical wet evergreen forests to subtropical montane forests to about 1,900m. **HABITAT** Common on trunks of wet evergreen forest plants. **FLOWERING PERIOD** October, April–August.

Lindley's Eria ■ *Eria lindleyi*

IDENTIFICATION Epiphytic orchid with a pseudobulbous stem rooting at base. Stems grow to about 15–45cm long, with leaves clustered at ends. Leaves sessile and 7.5–11cm long. Flowers white, about 1.2cm across. Two or three flowers per leafless inflorescence stalk (scape). Yellow floral bracts are very clear, and this feature can be used to identify the species. Fruit is a narrow, cleaved capsule, 2.5cm long. **DISTRIBUTION** Endemic. Submontane or mid-country tropical wet evergreen forests, extending to subtropical montane forests to 2,000m. **HABITAT** Common in trees at higher altitudes. **FLOWERING PERIOD** March, April, September–December.

Spectacular Eulophia
■ *Eulophia spectabilis*

IDENTIFICATION Ground orchid with a large tuberous rootstock. Stem very short. Normally two leaves per plant at the time of flowering. Leaves linear and long, about 30–60cm long. Raceme arises from rootstock. Peduncle about 40–60cm long and erect. Greenish-white flowers 1.2cm across. Veins of sepals purple. Lip white, or sometimes purple. Fruit is a spindle-shaped, ridged capsule about 5cm long. **DISTRIBUTION** Native. Tropical wet evergreen forests to 1,000m. **HABITAT** Rare species that grows in shade of trees of wet forests. **FLOWERING PERIOD** February–June.

Golden-Silver Weed ■ *Goodyera procera*

IDENTIFICATION
Herbaceous ground orchid that bears fragrant flowers. Stem of plant grows to about 50cm tall. Upper part of stem covered by leaf sheaths. Leaves large, 12–18cm long. Tiny flowers produced in terminal spike. Each flower 0.3cm across, and white or pink. **DISTRIBUTION** Native. Grows alongside grasses above 800m. **HABITAT** Rare species found in moist banks (close to tributaries) and hill cuttings in mid-hills. **FLOWERING PERIOD** January, February, June, September.

Pigeon Orchid ■ *Habenaria plantaginea*

IDENTIFICATION Small herbaceous ground orchid that grows to 40cm tall. Leaves very close to the ground, forming rosette-type leaf arrangement. Leaves oval in shape. Terminal inflorescence produces many pure white flowers with a characteristically trilobed lip. Middle lobe very narrow, and adjacent lobes broader. Flower about 1.6cm across, with green, slender, long (3cm) spur. **DISTRIBUTION** Native. Forests of intermediate zone and dry zone, but absent in arid areas. **HABITAT** Common under shade of trees of dry-zone and intermediate-zone locations. Rarely present in montane locations. **FLOWERING PERIOD** March–April, June.

Daffodil Orchid ■ *Ipsea speciosa*
(S: Nagameru ala)

IDENTIFICATION Orchid that is easily identifiable when walking in open patana grassland in the latter months of the year – its golden-yellow, showy flowers rise up over grasses during the flowering season. In the absence of flowers, the plant is difficult to

identify because the leaves are somewhat similar to leaves of grasses. This terrestrial orchid species has a pseudobulbous stem very close to the ground. One or two leaves are present at a time, each about 25–30cm long. Leafless lower stalk (scape) about 40cm long. Flowers large (about 5–6cm across). A few flowers may be in bloom at a time. Red or reddish-brown veins can be seen on lip. **DISTRIBUTION** Endemic. Montane-zone patana lands. **HABITAT** Endemic common plant that prefers to grow alongside grasses in open patana lands under cold climatic conditions. **FLOWERING PERIOD** September–March.

Luisia teretifolia

IDENTIFICATION Body structure of this epiphytic orchid very different in comparison to other orchid species. Stem and leaves alike – green, cylindrical and circular in cross-section (terete). Stem erect and about 15cm long. Leaves about 25cm long with rounded

apex. Flowering spikes tiny. Flowers very small, about 0.7cm across, and purple mixed with green. Lip deep purple and heart shaped. Another species in the same genus is *L. tenuifolia*. This differs from *L. teretifolia* by having slender long leaves, and not having a heart-shaped lip in the flower. **DISTRIBUTION** Native. Wet evergreen forests to 800m. Can also be seen on huge trees along roadsides. **HABITAT** Common on trunks and branches of evergreen trees. **FLOWERING PERIOD** March–June.

Oberonia wallie-silvae

IDENTIFICATION *Oberonia* is an orchid genus that has fleshy leaves, arranged in two vertical rows (distichous) and covering stem. This species grows on top of a host plant. Leaves decurved. An inflorescence or raceme comes out from plant axis. Inflorescence (10cm long) arises from base of plant, and top end of it droops. Flowers very tiny, about 0.2mm across. Flowering raceme appears light green at a distance. At close range, flower colour is reddish-brown. **DISTRIBUTION** Endemic. Tropical wet evergreen forests to 1,200m. **HABITAT** Very rare species that grows on tree trunks of wet and mid-country evergreen forests. **FLOWERING PERIOD** February–March, July.

Oberonia wightiana

IDENTIFICATION Small epiphyte with fleshy leaves about 3cm long. Leaves arise at base like those in a tussock, and are arranged in two vertical rows (distichous). Terminal raceme about 10cm long, slender and decurved. Raceme and tiny flowers yellow. **DISTRIBUTION** Native. Submontane or mid-country tropical wet evergreen forests at 548–2,134m. **HABITAT** Common on trunks of evergreen forest plants. **FLOWERING PERIOD** October, November.

Swamp Orchid ▪ *Phaius tancarvilleae*

IDENTIFICATION Giant herbaceous ground orchid with pseudobulbous stems near the ground. Each pseudobulb has 2–4 leaves, which are very large in comparison to leaves of other wild orchid species, at about 1m long x 15cm wide. Parallel veins prominent on leaf surface. Flowers large, about 6–11cm across. Flowers have multiple colours: sepals and petals red, orange-yellow mixed; lip orange-yellow to purplish-white. Column hidden inside tube formed by base of lip. Lip end open, with a wavy margin. **DISTRIBUTION** Native. Mid-country submontane forests at 900–2,100m. **HABITAT** Eye-catching orchid common under shade, and near seasonal waterfalls and tributaries of montane forest areas. **FLOWERING PERIOD** February, April–June, September–November.

Pale Pholidota ■ *Pholidota pallida*

IDENTIFICATION Epiphytic orchid with crowded pseudobulbous stems. Each pseudobulb about 1.5–6.5cm long and with a single leaf at a time. Leaves large, to about 30cm long, with parallel veins clear on blade. Flowering raceme drooping, with 10cm-long peduncle. Flowering portion about 5–12cm long. Many flowers per raceme. Individual flowers small (0.6cm across) and pinkish-white. Brown bracts very close to the flowers. Fruit is a globose capsule about 2cm long. **DISTRIBUTION** Native. Wet evergreen and submontane forests at 450–1,500m. **HABITAT** Common as epiphyte on tree trunks and rock surfaces in wet evergreen forests. **FLOWERING PERIOD** January, March, June–August.

Common Helmet Orchid ■ *Polystachya concreta*

IDENTIFICATION In this epiphytic species, lip is on top and column is below lip (in most orchid species the arrangement of lip and column is the other way around). Pseudobulbous

stems closed by sheaths of old leaves. Leaves alternate on stem and sheathed at base. Terminal panicle 15–40cm long together with peduncle. Flowers tiny, about 1cm across. Upside-down flowers have cap-like lateral sepals. Other floral parts hidden inside lateral sepals. Floral stalks green. Sepals and petals creamy-pink. Triangular, pointed floral bract per flower below flower stalk. **DISTRIBUTION** Native. Submontane or mid-country tropical wet evergreen forests to 1,700m. **HABITAT** Very common on trees of submontane or mid-country tropical wet evergreen forests. **FLOWERING PERIOD** March, April, July–October.

Green-flowered Pteroceras ■ *Pteroceras viridiflorum*

IDENTIFICATION A special characteristic of this epiphyte is its highly reduced stem.

Roots prominent and green. Leaves stalkless (sessile), about 5cm long, fleshy, with notched apex. Only midvein is clear. All leaves face one way. Spikes have 4–8 pairs of flowers at a time. Flowers 1.2cm across, bright pale green with white lip. **DISTRIBUTION** Native. Submontane or mid-country tropical wet evergreen forests. **HABITAT** Very rare, on trees in submontane or mid-country tropical wet evergreen forests. **FLOWERING PERIOD** February.

Fox-tail Orchid ■ *Rhynchostylis retusa*
(S: Hiwalwaliga)

IDENTIFICATION Beautiful epiphytic plant with woody stem about 25cm long. Leaves spread alternately, and grow as long as 15–37cm. Leaf tip divided into two unequal lobes – sometimes this division is not clear. Flowers borne in densely packed, drooping racemes 20–30cm long. Flowers each 1.2–2cm across, and white spotted with violet-pink. Lip tinged with dark pink. **DISTRIBUTION** Native. Tropical savannah in Uva and Eastern Provinces. **HABITAT** On trees close to streams in dry locations. **FLOWERING PERIOD** January, April, July, November.

Small-leaved Robiquetia ■ *Robiquetia brevifolia*

IDENTIFICATION Epiphytic orchid with a creeping habit. Tiny, drooping floral raceme is representative of the genus *Robiquetia*. It is reddish and tightly packed with tiny flowers

– a feature of this species. Stem lacks pseudobulbs. Thin, slender aerial roots visible on stem. Leaves arranged in two rows (distichous), coriaceous and unequally two lobed at apex. Leaves green or sometimes brownish-red. Floral raceme about 4cm long. Flowers tiny, about 1cm long x 3.5mm wide, and deep purplish-red with upperpart of lip yellow. Red tubular spur of similar length on floral peduncle. Fruit is a spindle-shaped (fusiform) capsule, 1.5 x 0.6cm. **DISTRIBUTION** Endemic. Submontane or mid-country tropical wet evergreen forests, extending to subtropical montane forests to 1,900m. **HABITAT** Prefers branches of trees in forests under cold climatic conditions. During the flowering season can be commonly seen when walking through roadside plants in places like the Knuckles and Horton Plains. **FLOWERING PERIOD** February–April, September, October.

Hyacinth Orchid ■ *Satyrium nepalense*

IDENTIFICATION Terrestrial ground orchid with basal leaves and an erect stem. Pink flower bunches are eye-catching when walking through wet patana grassland during the flowering season. Leaves hidden within grasses of its habitat. Many oblong tubers underground. Terminal flower spikes about 7–9cm long. Floral bracts shiny brown, equal in length to flower and recurved towards the ground when old. Sweet-scented flowers about 2.5–3.7cm across. Sepals, petals and lip bright pink. Pink tubular spur that is longer than flower peduncle. **DISTRIBUTION** Native. Wet patana land among grass above 1,200m. **HABITAT** Wet patana grassland. **FLOWERING PERIOD** July, September, October, November, December, January.

Golden-flowered Schoenorchis ■ *Schoenorchis chrysantha*

IDENTIFICATION Epiphytic, pendulous, climbing orchid with non-pseudobulbous stem. Pendulous habit and purple spots over stem distinguish this species from the other two *Schoenorchis* species. Stem very slender, with characteristic purple spots. Young areas of

stem ensheathed by petioles of older leaves. Leaves green, slender, long (about 12cm), pointed and with the appearance of stems. Raceme tiny, about 3cm long. Flowers yellow with orange veins, 3mm across. **DISTRIBUTION** Native. Subtropical montane forests to 2,200m. **HABITAT** Very common orchid growing on trees under cold climatic conditions. **FLOWERING PERIOD** March, May, June.

Philippine Ground Orchid ■ *Spathoglottis plicata*

IDENTIFICATION Introduced ornamental orchid that is a large herb growing to about 1m. Leaves very long (50cm), pleated near the ground. Flowers (2.5–3.5cm across) produced in terminal raceme 12–30cm long. Bract, flower stalk, sepals and petals bright pink. Young flower's lip-base yellow with brown spots, and hairy. Colour of flowers slightly variable according to the location. **DISTRIBUTION** Exotic. Montane, intermediate and dry zones. **HABITAT** Very common at high altitudes. Also grown as an ornamental plant in many parts of Sri Lanka. Has recently escaped and is becoming naturalized in the wild. **FLOWERING PERIOD** Throughout the year, but flowering peaks in February–August.

Chinese Spiranthes ■ *Spiranthes sinensis*

IDENTIFICATION Leafy ground orchid with fleshy, tuberous root fibres. Easily differentiated from other orchids by the spirally arranged terminal floral spike (genus name *Spiranthes* derives from this spiral habit). Leaves long (16–18cm) and linear, but difficult to notice because the plant grows with grasses. Leaf bases form sheaths around stem. Floral spike spiral, and densely covered by green, hairy (pubescent) bracts. Flowers white, tiny, 2.8-mm across, and spirally placed on spike. **DISTRIBUTION** Native. Wet patana grassland. **HABITAT** Common in wet patana grassland. **FLOWERING PERIOD** Throughout the year.

Spoon Vanda
■ *Vanda spathulata*

IDENTIFICATION Growth form that of an erect epiphyte. Stem non-pseudobulbous, erect and about 30cm long. Leaves green, coriaceous, arranged in two rows (distichous), and notched at tips (emarginated). During flowering, apical raceme grows out from middle area of stem. Flowers as large as 3.7cm across and yellow. They are flat, with sepals, petals and lip in a somewhat horizontal plane. 'Column' in the flower grows perpendicular to other parts of flower. Peduncles are long. **DISTRIBUTION** Native. Dry zone. **HABITAT** Very common species in huge trees of dry regions. **FLOWERING PERIOD** January, March–September, December.

Tessellated Vanda ■ *Vanda tessellata*

IDENTIFICATION Huge epiphytic *Vanda* species with non-pseudobulbous stem 30–60cm long, and simple, branching roots. Leaves long (15–20cm), strap shaped and coriaceous. Leaf apex has two unequal lobes. Leaves arranged in two vertical rows (distichous). Flower raceme grows from middle part of stem (usually from one of leaf bases). Raceme normally 15–20cm long. Flowers large, 5cm across, grey, greyish-blue, buff, red or yellow. Sepals and petals patterned. **DISTRIBUTION** Native. Tropical dry mixed evergreen forests in dry zone and along east coast. **HABITAT** Very common on trees in dry zone. **FLOWERING PERIOD** January, March–August, December.

Vanda ■ *Vanda testacea*

IDENTIFICATION Epiphytic orchid with non-pseudobulbous stems growing about 10–15cm long. This *Vanda* species can be easily identified by the colourful lip, which is purple, white or reddish-pink. Lips of flowers in the photograph are purple, but note the variability. Roots large, thick and vermiform (worm-like). Leaves similar in size, toothed at apex into

two lobes, and arranged in two rows (distichous). Midrib extends out and creates third lobe or tooth. Spikes grow out from leaf bases. Many flowers are in bloom on a spike at a given time. Flower 1.8cm across, yellow with a coloured lip. Fruit a long-pedicelled, club-shaped (clavate) capsule 2.5–3cm long. **DISTRIBUTION** Native. Tropical dry mixed evergreen forests and mid-country dry forests. **HABITAT** Very common on large roadside trees and jungle trees in dry locations. **FLOWERING PERIOD** February, March, September.

> **ASPHODELACEAE (ALOES)**
> This family mainly comprises herbs with succulent leaves in basal or terminal rosettes, though the plants also grow as climbers and pachycaul (thick-stemmed and sparsely branched) trees. There are about 800 aloe species in 15 genera. Aloes are found in the tropics, subtropics and temperate areas of the Old World, and have a preference for arid habitats. South Africa is a centre of diversity for the family, and aloes are absent (unless introduced) from South Asia.

Dianella Lily ▪ *Dianella ensifolia*
(S: Monarapetan)

IDENTIFICATION Similar to the ornamental *Dracaena* plants, but has green leaves without the colour patterns of *Dracaena* leaves. Stem is a rhizome. Leaves arranged in two rows (distichous), dark green and form a sheath at lower part of plant. Leaves as long as 30–60cm. Inflorescence is terminal with a long (about 75cm) ribbed stalk. Flowers open in the afternoon. Flower 1cm across with six perianth (sepals and petals) parts. Perianth segments light purple-blue with five or seven veins. Anthers bright yellow. Fruit is a very attractive berry, bluish-purple and shiny. **DISTRIBUTION** Native. Mid-country and montane regions to 1,200m. **HABITAT** Very common in home gardens, roadsides of tea plantations, abandoned places and similar areas. **FLOWERING PERIOD** April, July, August.

AMARYLLIDACEAE (DAFFODILS)

These perennial herbs are found worldwide, but mainly in the tropics and subtropics. The nearly 900 species in 50 genera have a centre of diversity in South Africa. Most species are found in seasonally dry areas (which may be surprising to those who associate daffodils with European woodland and the poetry of William Wordsworth). Daffodils can be ground dwelling, aquatic or epiphytic. They bear tubular flowers in an inflorescence, with each flower being borne on a single stalk. The flowers are usually made up of six distinct or fused petaloids. Daffodils are economically important as ornamental plants because of their spectacular flowers. Some species are reputed to have anticancer alkaloids.

Amazon Lily ▪ *Eucharis grandiflora*

IDENTIFICATION Stem of this perennial is an underground bulb that looks like a large onion. Leaves deep green, large (nearly 40cm long x 20cm wide), stalked and spectacular. Flowers large, showy and borne in umbel of 3–10 flowers on erect stalk that arises from the underground stem. Sweet scent of the flowers can be detected from a distance, and plant can be identified by smell with experience. Flowers have six petals, and after a shower beautiful water droplets can be seen on the petals. Flower has prominent central cup (corona), streaked or tinged with green. White anthers fused with corona. Style free and extends over stamens.

DISTRIBUTION Exotic. A hybrid plant, found in ground vegetation of lower elevations in wet- and montane-zone locations. Also commonly grown as an ornamental in home gardens. **HABITAT** Prefers moderately cold, shaded places. Commonly grows with ground vegetation, even in shady areas. **FLOWERING PERIOD** No exact time duration for flowering. Normally produces flowers twice a year.

Barbados Lily ■ *Hippeastrum puniceum*

IDENTIFICATION Bulbous perennial in which leaves and flower stalks arise from bulb. Leaves green, succulent, long (nearly 40cm), narrow (nearly 3cm wide) and tapering towards end. They are not prominent when plant is in flower. Flower stalk a green, erect, long (nearly 45–60cm), hollow tube. Flowers borne in an umbel on flower stalk. Umbel has two green, lanceolate-shaped bracts at base. Flowers large, showy and reddish to orange in colour, bending and opening towards the ground. No distinct sepals and petals. Six tepals in each flower, and tepal base tinted white. Underside of tepals striped with green. Village children use these flowers to make simple, temporary flutes. **DISTRIBUTION** Cultivated. Abundant in dry-zone lowlands. Also found in drier aspects of montane zone. **HABITAT** Open, dry places with abundant sunlight, but will tolerate modest degree of shade. **FLOWERING PERIOD** Early months of the year.

ERIOCAULACEAE (PIPEWORTS)
Pipeworts are annual or perennial herbs found mainly in the tropics and subtropics. The greatest species diversity is in South America and Africa, with a few species also found in North America and Europe. The flowers are characteristically button shaped on leafless, long peduncles. Some species may bear the flowers in an umbel with up to 100 flowering heads. Each head can, in turn, have many flowers. The petals are usually fused with the free sepals, and the flowers are radially symmetric. The leaves form a spirally arranged (rarely distichous) dense rosette at the base, and are grass-like. In aquatic species the leaves are filiform (thread-like). A few species are popular as potted plants.

Pipewort ■ *Eriocaulon brownianum*

IDENTIFICATION Grows as a tussock. Leaves (60cm long) grass-like, linear, long, acuminate and erect. Flower stalk (or scape) 30–80cm long. Inflorescence is a densely packed head of stalkless flowers (a capitulum), which is hemispherical and becomes spherical (globose) at maturity. White flowers 1–2cm wide. Flower head like a button – a common name for this plant is Ladies' Button Plant. Tallest *Eriocaulon* species in Sri Lanka. **DISTRIBUTION** Native. Upper montane locations to 1,000–2,000m. **HABITAT** Swampy grassland of upper montane locations. **FLOWERING PERIOD** Photographed in November. There is little published information on the flowering period.

CYPERACEAE (SEDGES)
The sedges look superficially like grasses, and in temperate and subarctic regions they can become the dominant plants in damp or marshy areas. They occur worldwide and are absent only from Antarctica. There are about 4,000 species in 70 genera, and they are mainly perennials, but a few are annuals. They grow as shrubs, rarely as lianas. Their stems are trigonous (three angled) in cross-section. The ripe (mature) fruits are usually achenes (single dry seeds that do not split open), or rarely drupes (fleshy fruits with many seeds). *Cyperus papyrus* gave its name to 'paper'. The stems of various species are used in thatching and basket making.

Jointed Schoenoplecius ■ *Schoenoplecius articulatus*

IDENTIFICATION Perennial herb that forms tussock-like, large clumps with a fibrous root system. Stems (known as culms) erect (sometimes recurved), hollow, smooth, shiny and dark green. Flower head (bunch of flowers) 2–4cm in diameter, located midway on a culm. Each flower head has more than 60 densely arranged spikelets. Young spikelets creamy, tinged with light green and brown. **DISTRIBUTION** Native. Low altitudes of lowland dry zone. **HABITAT** Marshy places at low altitudes; often seen in shallow water along margins of ponds. **FLOWERING PERIOD** Early months of the year.

> **COMMELINACEAE (SPIDERWORTS)**
> This family of herbs and climbers and a few epiphytes is found in the warm temperate
> tropics and subtropics. Some species are annual. Spiderworts occur in forests and
> grassland. In the temperate zone they are richest in North America and Asia, and
> their species diversity is greatest in tropical areas; they do not occur in Europe. There
> are about 650 species in 40 genera. The plants typically have swollen nodes and leaves
> that are spirally arranged, or distichous. The flowers grow on inflorescences at the end
> of a shoot, or from a leaf axil, and are usually vertically symmetrical, though flowers of
> a few species have radial symmetry. They are strongly scented in some genera (such as
> *Palisota*, *Tripogandra*), but without nectar.

Benghal Dayflower ▪ *Commelina benghalensis*
(S: Girapala)

IDENTIFICATION Annual climbing herb. The plant spreads using aerial roots above the
ground that root themselves into soil at nodes (stoloniferous). Leaves arranged in two
vertical rows (distichous), sessile, and attached to stem with a sheath. Leaf ovate or ovate-
elliptic, with acute apex. Flowers grow on a spathe and have three petals. Lower petal
smaller than the two laterals. Flowers blue, each with six anthers. Three yellow anthers are
fertile, three infertile (staminoides). They are, respectively, on the upper and lower parts of
the flowers, and the fertile anthers have a yellow powdery appearance. The anthers being of
two different types is known as anther dimorphism. White hairs at bases of petals, anthers
and ovary. **DISTRIBUTION** Native. Throughout Sri Lanka. **HABITAT** Marshy land, weedy
places and roadsides. **FLOWERING PERIOD** October–February, April, May and July.

PONTEDERIACEAE (WATER HYACINTHS)
This family of aquatic plants is found in tropical and subtropical zones, and comprises more than 30 species in nine genera. The species grow in many forms; some are emergent, some flow and some are submerged. The inflorescence is a terminal spike, often with showy flowers that are insect pollinated. The leaves are distichous (arranged in two vertical rows), and usually elongated. The vertically symmetrical flowers are bisexual and short lived, typically living for just a day. In some genera, such as *Eichhornia*, the floral axis dips below the water's surface and the fruits mature underwater. Several species are used as aquatic ornamentals, and the family also includes plants such as the Water Hyacinth E. *crassipes*, which are highly invasive and blank out bodies of water. The succulent leaves of some species are used as food for livestock, and the stems of Water Hyacinths are used for basket making.

Pickerel Weed ▪ *Monochoria vaginalis*
(S: Diyahabarala, Jabara)

IDENTIFICATION Plant with an erect shoot arising from a creeping rhizomatous stem hidden under a muddy bottom. Leaves simple, coriaceous, with a long, spongy, sheathing petiole. Upper surface of leaf blade dark green; paler underneath. Terminal flower bunches contain 3–16 open flowers at a time. Six tepals per flower, lavender in colour and arranged in two rows. Stamens conspicuous, with yellow anthers. Leaves eaten as a vegetable.
DISTRIBUTION Native. Lowlands. **HABITAT** Common in marshy places such as abandoned paddy fields, ditches and edges of tanks. **FLOWERING PERIOD** Throughout the year.

> **ZINGIBERACEAE (GINGERS)**
> The gingers are herbaceous perennials that include many economically important plants such as the Ginger *Zingiber officinale*, Cardamom *Eletarria cardamomum* and Turmeric *Curcuma longa*. The family is rich in aromatic volatile oils, which are used in food, perfumes and medicines. The plants have creeping horizontal or tuberous rhizomes, and a few species are epiphytic. Their centre of diversity is in Indo-Malaysia, and they are found mainly in the tropics. There are about 1,300 species in around 50 genera. Their simple leaves are arranged in two vertical rows. The vertically symmetrical flowers are showy, and the inflorescences can arise from a leaf shoot or directly from the rhizome.

White-flowered Ginger ■ *Curcuma albifora*
(S: Harankaha)

IDENTIFICATION Leaves arise from underground stems (rhizomes). A 'false stem' or 'pseudo stem' is formed by leaves appearing above the ground. Leaves nearly 30–50cm long and petiole about 10–20cm in length. Leaf blade oblong in shape, with acute tip. Characteristic brownish-red colouration along midrib of leaf – a feature that makes identification of this plant easy. Flowering spike, sheathed at base, produced when young leaves are developing. Both flower bearing and non-flower bearing bracts are borne on spike. Bracts in basal area green in colour, and most are flower bearing. All other bracts reddish-purple. Flowers white, tube-like and lobed. Floral tube about 3cm long. Labellum or lower floral lobe yellow. **DISTRIBUTION** Endemic. The photographs were taken in Nawalapitiya. Recorded in Kithulgala (Kelani Valley Forest Reserve), Bopath Ella and Maskeliya. **HABITAT** Rare endemic that can be seen in undergrowth vegetation in wet montane forest areas, and along riversides and water channels in montane localities. **FLOWERING PERIOD** Early part of the year, from March to June.

BERBERIDACEAE (BARBERRIES)
There are about 700 species of barberry in around 16 genera, and the family comprises herbs and shrubs. The temperate regions are home to the herbaceous species. The family includes the woody shrubs in the genus *Mahonia*, grown in temperate gardens around the world. Barberries typically have berry-like fruits and spiny stems, and the plants are widely cultivated for their ornamental value. The berries on the *Mahonia* shrubs in Victoria Park in Nuwara Eliya are a magnet for wintering Pied Thrushes. The leaves are variable and vary from simple, to pinnate to compound, and the bisexual, radially symmetric flowers are borne on panicles, racemes or peduncles.

Barberry ▪ *Berberis ceylanica*
(S: Daruharidra)

IDENTIFICATION Shrubby plant that grows to 3m tall. Stem brown and grooved. Deep green, hard, shiny cluster of leaves arises from each node. Leaves ovate, leaf tip mucronate. A few reddish-brown spines below each leaf cluster. An inflorescence with many flowers (about 15) arises from each node. Flowers (1.5cm diameter) bright yellow and showy. Ovary red and easily seen by opening the petals. Both sepals and petals yellow, with sepals tinged reddish-brown underneath. Peduncles reddish-brown. Fruit is an obovoid berry (1.5cm long). Young fruits pinkish-red, mature ones purple. Stigma remains on fruit as a cap. **DISTRIBUTION** Endemic. Montane forests and forest borders, to about 2,200m. Areas from central hills, especially Horton Plains National Park and its surrounds. **HABITAT** Roadsides and wet patana grassland. Prefers to grow as an undergrowth plant in montane forests. **FLOWERING PERIOD** Early months of the year.

> ### RANUNCULACEAE (BUTTERCUPS)
> This large family comprises more than 1,800 species worldwide, with the majority growing in cold regions. All the plants are herbaceous, except for the genus *Clematis*, which is woody. Species in some genera are poisonous, and those in the New World have a greater variety of colours than those in the Old World. The leaf form is highly variable, with some species having simple leaves, others deeply lobed leaves and some thin, feathery leaves.

Buttercup ▪ *Ranunculus sagittifolius*

IDENTIFICATION Perennial herb with a rhizomatous stem from which the leaves arise. Eye-catching, bright yellow, waxy petals enable easy identification of the plant – this is very easy during the flowering season because the flowers bloom above the grasses with which the species associates. Petiole sheathed and hairy. Leaf long (about 10cm), cordate, with an obtuse tip. Leaf base has two rounded lobes. Reticulate venation very clear. Inflorescence tall (about 90cm). Flowers 2–3cm across. Sepals green and boat shaped. Five petals, spreading, overlapping, and forming cup. Filaments very short with yellow anthers. Many green styles with yellow beak at centre of flower. **DISTRIBUTION** Endemic. Can be seen in upper montane zone and upper montane wet patana grassland. **HABITAT** Wet marshy places in between grasses or sedges. **FLOWERING PERIOD** April–October.

NELUMBONACEAE (LOTUSES)

This family of aquatic herbs contains a single genus, *Nelumbo*, which was until recently included in the family Nymphaeceae – the water lilies. However, molecular phylogenetics have resulted in the genus being assigned to its own family with what are known as the lower eudicots. There are two species in the family, *Nelumbo lutea* and *N. nucifera*, although some authors believe that the latter is a subspecies. The plant has religious importance in many Asian countries, and its edible tubers, young leaves and seeds are eaten.

Lotus ▪ *Nelumbo nucifera*
(S: Nelum; T: Tamarai)

IDENTIFICATION Aquatic plant with rhizomatous stem. Leaves flat and circular, about 50–60cm across. They float on the water's surface, and sometimes rise above the water on a rigid petiole. Leaf surface dark green; light green underneath. Veins prominent on lower surface. About 20 veins start from petiole point and radiate towards leaf margin. Prickles on petiole. Flowers large, about 20cm across, with many tepals. Outer tepals larger than middle ones. Colours of tepals vary, and may be pink, purplish-pink, pinkish-white or white.

Many stamens. Numerous stigmas are sticky, yellow and joined together, making a disc at centre of flower. Middle disc makes a fruit. Ripe fruit spongy, black and light, and can float on the water's surface after being detached from plant. Cavities of fruit enlarge and make room for seeds (or nuts) to fall away. **DISTRIBUTION** Native. Abundant in dry zone. Also cultivated in many places as an ornamental plant. **HABITAT** Aquatic habitats such as ponds and lakes. **FLOWERING PERIOD** Throughout the year.

DILLIENIACEAE (SIMPOHS)
This family is found in tropical and subtropical regions, and contains about 450 species in 11 genera. It is widely distributed in Australia, and the genus *Hibbertia*, popular as a garden plant, ranges from Madagascar to Australia. Most species are woody, growing as shrubs, trees and lianas, and some have edible fruits. The flowers are showy, radially symmetric, and have 3–5 petals with numerous stamens. The leaves are simple and alternate, often with prominent veins and dentate margins. The timber from *Dillenia* is used in boat building.

Acrotrema uniflorum

IDENTIFICATION Endemic perennial herb with woody rootstock and short stem. Simple leaves large (10cm long), oval to obovate shaped, and leaf arrangement is a rosette. Leaf tip obtuse, with base having ear-like lobes (auriculate), and petiole sheathed. Leaf blade hard and brownish-green. Midrib and secondary veins very clear on leaf blade. Dominant hairs along leaf margin characteristic in this species. Inflorescence is a type of raceme. One, two or rarely many flowers per inflorescence. Flowers small (nearly 2cm across), bright yellow. Five ovate petals per flower, and many yellow anthers at centre of flower. Three white styles grow over stamens. **DISTRIBUTION** Endemic. Wet evergreen forests at 500–1,300m. **HABITAT** Moist, shady underground habitats of montane forests. **FLOWERING PERIOD** February–April, August–September.

Water Simpoh ■ *Dillenia suffruticosa*

IDENTIFICATION Small, spreading tree that grows to 3m tall. Ovate leaves large (30cm long), simple, deep green above and pale green below. Petioles grooved. Flowers bright yellow, showy and large (8–12cm across). Each flower has five petals and five sepals. White stamens and yellow staminodes (sterile stamens not bearing pollen) clustered at centre of flower. Fruits reddish-brown and spherical (globose). Mature fruits split (dehiscent), opening up by spreading of sepals and unfolding of carpels. Opened fruits appear like star-shaped flowers. Carpels bright scarlet. **DISTRIBUTION** Exotic. Wet lowlands to 900m. **HABITAT** Ditches of rice fields, neglected plantations, roadsides and similar places. The plant is becoming invasive. **FLOWERING PERIOD** Throughout the year.

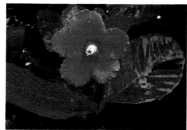

Kekiriwara ■ *Schumacheria castaneifolia*
(S: Kekiriwara)

IDENTIFICATION Large shrub, sometimes growing as tree to 5m tall. Leaves simple, ovate, large (12–20cm long) and deep green. Leaf apex acute. Secondary veins parallel to each other and very clear. Inflorescences are terminal panicles. Spikelets of panicle are drooping, and panicles are many flowered. Yellow flowers small (10–14cm across). **DISTRIBUTION** Endemic. Wet lowlands to 100–800m. **HABITAT** Main habitat is primary rainforests, but has recently spread to secondary rainforests. **FLOWERING PERIOD** Throughout the year.

OXALIDACEAE (WOOD SORRELS)
This family of about 800 species is found mainly in tropical and subtropical regions, and comprises annual or perennial herbs, climbers, shrubs and small trees. Despite the large number of species, they are all placed in five or six genera (depending on the author). The family includes the Starfruit *Averrhoa carambola*, which is edible, although it is dominated by weedy species in the genus *Oxalis*. Some species grow in clumps, spreading on stolons. The flowers have five sepals and five petals, but the inflorescences are variable from species to species. The sexual arrangement also varies, with some species being unisexual and others bisexual.

Yellow Woodsorrel ■ *Oxalis corniculata*
(S: Embul embiliya, Heen embul embiliya, Rata embala; T: Puliyari)

IDENTIFICATION Very small herbaceous runner – can run about 15cm across the ground. Stem very slender and green or purple. Petioles long (about 10cm), purple or green. Compound leaves have three obcordate leaflets, deeply notched (emarginate) at tips. Floral peduncle 12cm long and erect. Flower very tiny (1cm across), yellow. Petals spoon shaped (spathulate), rounded at apex. **DISTRIBUTION** Exotic. Found throughout Sri Lanka. **HABITAT** Cultivated and disturbed areas. **FLOWERING PERIOD** Throughout the year.

Pink Woodsorrel ▪ *Oxalis debilis* var. *corymbosa*
(S: Embul embiliya)

IDENTIFICATION Plant looks stemless. Underground bulbous stem can be seen when uprooted. Leaf petioles arise directly from bulb. Leaves compound with three sessile leaflets, tip of each of which has a notch. Lobes of leaflets rounded. Flower 1.5cm across. Petals purplish-pink with pale green at base. Clearly marked purple veins along petals. **DISTRIBUTION** Exotic. Mid-country and sometimes higher elevations. **HABITAT** Cultivated land and disturbed open places. **FLOWERING PERIOD** Throughout the year.

Broadleaf Woodsorrel ▪ *Oxalis latifolia*
(S: Embul embiliya)

IDENTIFICATION Growth is from a bulbous stem, as is common in other *Oxalis* species. Leaves compound with three leaflets that are nearly triangular and notched at tips. Flowers (1.4cm across) in an umbel with many flowers – usually one can be seen at a time in an umbel. Five petals slightly obovate, purple with green base. Dark green markings (veins) clear on petal base. **DISTRIBUTION** Exotic. Mid-country and higher elevations. **HABITAT** Cultivated and disturbed areas. **FLOWERING PERIOD** Periodical flowering throughout the year.

ELAEOCARPACEAE (OIL-FRUITS)

This mainly tropical and subtropical family of herbs, trees (including some important for their timber), shrubs and ericoid shrubs is absent from continental Africa and North America. There are about 600 species in 15 genera. The largest genus, *Eleaocarpus*, is found in East Asia, Australia, Indo-Malaysia and the Pacific area. Many species have narrow distributions. In Sri Lanka, four endemic *Eleaocarpus* species have evolved in the cloud forests. A few species produce edible fruits. The leaf arrangement can vary, and may be alternate, opposite, or in whorls or spiral arrangements. The leaves are generally simple, with entire or serrate margins, and the flowers have 3–5 sepals and 4–5 petals.

Mountain Oil-fruit ▪ *Eleaocarpus montanus*

IDENTIFICATION Grows as a tree. Young parts of stem hairy (pubescent). Leaves oval shaped, coriaceous, 5–7cm long. Racemes 4–5cm long, erect (mostly), with nearly 20 flowers per raceme. All flowers attached to raceme, so that they face one side. Five sepals

scarlet and harder than petals. Five petals white and broadly triangular in shape. Petal tips deeply divided, giving a characteristic feather-like appearance. Both surfaces of petals hairy. Many anthers at centre of flower, white in young flowers, pale brown in old ones. Fruits are spherical (globular), about 2.3–2.5cm long, minutely velvety and tinged red-brown. **DISTRIBUTION** Endemic. Upper montane forests and other locations at 1,900–3,000m. **HABITAT** Restricted to upper montane forests. **FLOWERING PERIOD** January–April, September.

> **ERYTHROXYLACEAE (COCA)**
> There are around 250 species in four genera in this pantropical family. The plants grow as both trees and shrubs, and may be deciduous or evergreen. The small flowers are radially symmetric – in some species they are unisexual, but more frequently they are bisexual. They have five sepals, five petals, and 10 stamens in two whorls of five. The leaves of many species contain alkaloids, with the most famous plants being Coca *Erythroxylum coca* and *E. novagranatense*, which are used in medicine and to manufacture the addictive drug cocaine.

Coca ■ *Erythroxylum coca*

IDENTIFICATION Grows as a shrub. Flowers solitary in leaf axils. Flower 1cm across, with five sepals, white on upper surface and greenish beneath. Five petals smaller than sepals, white. The 10 stamens are united with the petals, and appear to lie on top of them (epipetalous). Ripened fruits are upright, scarlet and one seeded. Sepals persistent on fruits.

DISTRIBUTION Exotic. Botanical gardens, and occasionally home gardens. **HABITAT** Introduced and cultivated in different parts of Sri Lanka. **FLOWERING PERIOD** June, July.

Red Cedar ■ *Erythroxylum monogynum*
(S: Agil; T: Devadaram, Chemmanatii)

IDENTIFICATION Much-branched shrubs or small trees, 2–4.5m tall. Leaves simple, entire, obovate and alternately arranged on stem. Flowers white, 1cm across; 1–4 flowers together in leaf axils. Ten stamens with a staminal tube. Fruits less than 1cm long, scarlet when ripe. Calyx persistent on fruits. One of the familiar trees for those going on game drives in the national parks of the dry lowlands. **DISTRIBUTION** Native. Very common in dry zone, to 600m. **HABITAT** Dry-zone forests. **FLOWERING PERIOD** August–February.

Bo-kera ▪ *Gomphia serrata*
(S: Bo-kera; T: Katharai)

IDENTIFICATION Slim tree to 25m tall. Leaves about 12cm long, with a shiny surface, and thin and stiff, like paper (chartaceous). Leaf margin finely toothed (denticulate). Inflorescences many flowered and 8–12cm long. Flowers 1cm across. Five reddish-brown sepals and five yellow petals. Fruit about 1cm long, with persistent sepals. **DISTRIBUTION** Native. Wet-zone and low montane forests. **HABITAT** Wet-zone localities, slippery rocks. **FLOWERING PERIOD** January–May.

CLUSIACEAE (MANGOSTEENS)

This is an economically important, largely pantropical family that grows as herbs (annual or perennial), shrubs and trees. The plants provide fruits, latex, essential oils, drugs, pigments and resins. The leaves are simple, usually without stipules and generally opposite, but sometimes alternate or whorled. The radially symmetric flowers can be unisexual, or bisexual in different species. They are solitary, or borne in a cyme at the end of a shoot or in axils. Mangosteen fruits can be a fleshy, or dry berries or drupes. *Callophyllum* and *Mesua* species are important for their timber. *Garcinia* species have edible fruits, including the mangosteen.

Beach Calophyllum ▪ *Calophyllum inophyllum*
(S: Domba; T: Punai, Punnaigam)

IDENTIFICATION Smooth (glabrous) tree that grows to 20m or taller. Mature tree has deeply fissured, grey to blackish bark; inner living bark pink with clear yellow exudate. Leaves simple, hard and densely crowded on twigs. Leaf shape elliptic to obovate-elliptic. Leaf about 10–20cm long, with prominent middle vein in lower surface. White flowers (3cm in diameter) borne in axillary racemes. In each flower centre are crowded yellow stamens. Stalks of raceme, pedicels and flower buds cream in colour. Fruits round (globose), yellow or greenish, to 4cm in diameter (usually 2–3cm). Flesh around fruit (pericarp) pulpy and astringent. **DISTRIBUTION** Native. Lowlands along coast. Sometimes in sandy areas in dry-zone forests. **HABITAT** Sandy and rocky seashores with high temperatures. **FLOWERING PERIOD** April, May, June.

Pitch Apple ■ *Clusia rosea*
(S: Gal Goraka)

IDENTIFICATION Tree-like, huge shrub, typically to 3m tall, and providing dense cover over the ground. Bark smooth and mostly ash coloured. Leaf arrangement opposite. Leaves deep green, shiny, large (14cm long), thick, coriaceous and ovate. Leaf base tapers towards petiole, leaf margin smooth. Middle vein of leaves clear, other veins not clearly visible. Few (mostly three) showy pink flowers produced at tips of branchlets. Five petals with clear dark pink veins. Fruit is a spherical capsule. Reddish-brown sepals persistent on capsule base. Deep reddish-brown, flower-like stigma persistent on top of green and greenish-red

capsule. Mature capsule opens by splitting along lines of carpal junctions (septicidal dehiscence). Capsule white inside, with winged axis in centre. Seeds joined to capsule's central axis (columella) with gummy brown flesh. **DISTRIBUTION** Exotic. Wet rocks in montane and lowland wet forests. **HABITAT** Common in wet, rocky places with moderate temperatures. Often occurs along streams, in marshy areas of mid-hills and hill-zone forests, and in disturbed areas associated with some cultivation, such as those utilized for tea or coffee. **FLOWERING PERIOD** Photographed in March. Information on flowering period absent in books consulted.

Ceylon Ironwood ■ *Mesua ferrea*
(S: Diya na)

IDENTIFICATION Tree that normally grows to about 10m tall. Smooth brown bark. Leaves thick, long and tapering towards acuminate tip. Young leaves colourful and very attractive, pink to red and drooping. Large flowers scented and showy. Each flower has four white petals; numerous yellow or orange anthers clustered in centre. **DISTRIBUTION** Endemic. Common in lowlands and mid-elevation areas, sometimes appearing as mono-species forests (Na forests). Also grown as ornamental plant in botanical gardens and religious places. **HABITAT** Slow-growing species common in moist places such as marshes and stream sides with good sunlight. **FLOWERING PERIOD** Early months of the year.

Hypericaceae (St John's-Worts)
Until recently this family was treated as one of the subfamilies in the mangosteens (Clusiaceae). The flowers are bisexual and many of the species show heterostyly – an evolutionary adaptation in which the flowers bear stamens of two or more different lengths to help improve pollination success. The stamens are typically in bundles.

Mysore St John's Wort ■ *Hypericum mysurense*

IDENTIFICATION Much-branched shrub 1–2 (–3)m tall. Leaves stiff, very closely arranged in four ranks. Flowers 4–5cm across, in few-flowered terminal cymes. Many yellow stamens. Ovary light green, and very clear at centre of flower. **DISTRIBUTION** Native. Hill-country forests above 1,500m. **HABITAT** Very common in wet patana and bushy places in hill-country forests. **FLOWERING PERIOD** Throughout the year.

PASSIFLORACEAE (PASSION FLOWERS)
Plants in this family are mainly tropical climbers (woody and herbaceous) with
tendrils, such as the familiar Passion Fruit *Passiflora edulis*. Some are woody shrubs and
a few grow as trees, and about 700 species in around 16 genera occur in the tropics
and subtropics. Although the well-known Passion Fruit is edible, the family contains
many species that are toxic. Some are used as host plants by butterflies that are able to
absorb the alkaloids, which make them distasteful to predators. The genus *Passiflora*
has its centre of diversity in the humid regions of South America, with more than 500
species. The radially symmetric flowers are showy.

Red Banana Passionfruit ▪ *Passiflora antioquiensis*

IDENTIFICATION Plants in the genus *Passiflora* are climbers. Stem has dense covering

of short, soft hairs (tomentose). Leaves
simple. Flowers solitary, red to pink-red,
10–14cm across, pendulous on pedicel
20–25cm long. Petals and sepals alike.
Floral centre white. Corona filaments
short and purple. Stalk carrying both
stamens and ovary hangs out from where
petals are joined (an androgynophore).
Fruit longitudinally ribbed, 12cm long.
DISTRIBUTION Naturalized exotic
species. Montane zone to 2,200m.
HABITAT Prefers cold climatic
conditions. **FLOWERING PERIOD**
October–January.

Goat Scented Passion Flower ▪ *Passiflora foetida*

IDENTIFICATION Grows as herbaceous climber to about 3m long. Has an extremely unpleasant odour, hence the name *P. foetida*. Stem crowded with white hairs. Leaves simple, flat, nearly circular (suborbicular), and usually three lobed. Flowers 2.5–5cm in diameter, pink, lilac, purplish, or sometimes white. Sepals and petals alike, but petals slightly shorter than petals. Corona filaments are in several series. Filaments of outer series similar in length to petals. Those at base and centre of flower are purple. Androgynophore shorter than in other *Passiflora* species. Filaments flattened. Fruit is a dry berry enveloped by persistent involucre. **DISTRIBUTION** Naturalized exotic species. Dry zone. **HABITAT** Very common on waste ground. **FLOWERING PERIOD** August–May.

Sweet Granadilla ▪ *Passiflora ligularis*

IDENTIFICATION Climber with tendrils arising from leaf axils. Highly attractive white and reddish-purple banded filaments of corona help with easy identification of the species. Stem rounded. Leaves large (15cm long), simple, broadly ovate, tip acuminate, base cordate. White or pale pink flowers 6–9cm in diameter, solitary or in pairs, arise from leaf axils. Petals cream in colour. Filaments of outer rows of corona as long as petals, radially symmetric, circular in cross-section (terete), blue at apex and banded with white, and reddish-purple below. Filaments and anthers flattened. Ripe fruit ovoid and edible, 6–8 x 4–5cm, yellow or purple. **DISTRIBUTION** Naturalized exotic species. Montane region to 2,000m. **HABITAT** Montane locations. **FLOWERING PERIOD** April, October–December.

Banana Passion Flower ■ *Passiflora mollissima*

IDENTIFICATION Climbing plant, easily identified by large, pink pendulous flowers and ellipsoid fruits. Leaves simple, very clearly lobed into three. Flowers axillary and solitary. Peduncles 2–6cm long. Petals and sepals light pink. Corona not conspicuous, unlike in other *Passiflora* species, and appears as purple ring. Androgynophore as long as 8–12cm. Stamens longer than styles. Fruit ellipsoid or oblong-ovoid, 7–12cm long. **DISTRIBUTION** Exotic. Montane forests at 1,500–2,000m. **HABITAT** Montane forest margins and cleared forest areas. **FLOWERING PERIOD** February, October–December.

Giant Granadilla ▪ *Passiflora quadrangularis*
(S: Tun-Tun, Desi-Puhul)

IDENTIFICATION As implied in the scientific name, shape of stem is quadrangular – this feature and the attractive corona help with identification. Four-angled, winged climbing stem. Leaves simple, entire, broadly ovate. Midrib and secondary veins very clear. Flowers 10–12cm in diameter, whitish or pinkish. Upper surface of sepal reddish-pink, underneath green. Petals deep pink. Outer filaments of corona thread-like, almost equal in length, banded with reddish-purple and mottled with pink-blue. Androgynophore very short. Filaments and styles purplish-pink. Mature fruit very large (about 30cm long) relative to fruits of other passion plants. Fruit shape oblong-ovoid. Fruit is edible. **DISTRIBUTION** Cultivated exotic plant. Wet lowlands and mid-elevations. **HABITAT** Cultivated land, forest edges. **FLOWERING PERIOD** September–December.

Corkystem Passion Flower ■ *Passiflora suberosa*
(S: Veldodan)

IDENTIFICATION Climber with tendrils. May bear two types of leaf, entire or lobed. Flowers small, 1–2cm in diameter, solitary or in pairs in axils of leaves. Absence of petals is a special characteristic of this species. Sepals pale greenish-yellow. Corona filaments in two series. Filaments in outer series greenish-yellow and curve down; filaments in inner series tightly packed and purplish-brown. Androgynophore tinted with purplish-brown spots. Fruit is a 1.5cm-long, round (globose) berry. **DISTRIBUTION** Naturalized exotic species. Throughout Sri Lanka to 2,000m. **HABITAT** Forests and forest margins. **FLOWERING PERIOD** June–December.

Climbing Flax ■ *Hugonia mystax*
(S: Maha getiya, Bu getiya, Watti weti)

IDENTIFICATION Climbing shrub to about 3–4m tall. Branches spreading. Leaves crowded at ends of twigs, simple, oval and alternate. Flowers terminal in upper axils, rather large, 2.5–3.75cm in diameter. Five sepals, five petals, five styles and 10 stamens. Yellow anthers. Bright yellow petals. Fruit drupaceous (stony seed set in a fleshy covering), round, 1cm across, bearing persistent sepals. Ripe fruit scarlet/red. **DISTRIBUTION** Native. Lowland dry zone. **HABITAT** Forest edges and scrub forests in dry zone. **FLOWERING PERIOD** After monsoon, but can be seen throughout the year.

Red-bead Tree ■ *Adenanthera pavonina*
(S: Madathiya; T: Anaikuntamani)

IDENTIFICATION Tree that grows to about 20m tall. Compound leaves bipinnate with 2–6 pairs of leaflets (pinnae). Flowers minute, numerous in each inflorescence, and

creamy-white at first, later becoming yellow. Fruits are long, narrow pods, curved or slightly twisted. Seeds bright red with black caps. They are non-edible, and are collected by people for decorative use. **DISTRIBUTION** Cultivated. Low country to 1,300m. **HABITAT** Common on cultivated land. **FLOWERING PERIOD** Throughout the year.

Matara Tea ■ *Cassia auriculata*
(S: Ranawara; T: Avarai)

IDENTIFICATION Shrub-like tree that grows to about 3m tall. Occurs mainly as a shrub at low elevations. Leaves compound with 8–12 pairs of leaflets. Stipules large with ear-like (auricle) shape, at base of petiole. A few flowers are aggregated into large, terminal panicle in axils of upper leaves. Flowers large, bright yellow and attractive, with a shiny appearance. Fruit is a legume with 12–20 seeds inside it. **DISTRIBUTION** Native. Low-country dry zone. **HABITAT** Commonly found in warm, moist locations. **FLOWERING PERIOD** July, August.

Peanut Butter Cassia ■ *Cassia didymobotrya*
(S: At tora, Rata tora)

IDENTIFICATION Shrubby plant that grows to about 4.5m tall. Leaves long and compound, and 7–18 pairs of leaflets per compound leaf. Leaflets elliptic-oblong with stiff tips (mucronate apex). Inflorescences are racemose, axillary, narrow, to about 40cm long. Bracts large, hairy, and dark brown to black. Flowers yellow and showy. **DISTRIBUTION** Exotic. Throughout Sri Lanka. **HABITAT** Terrestrial moist places, for example near riverbanks, paddy fields and abandoned areas. **FLOWERING PERIOD** March.

Showers of Gold ■ *Cassia fistula*
(S: Ehela, Ekala; T: Tirukkontai, Kavani, Konnei)

IDENTIFICATION Typically grows as tree to 10m tall. Leaves long, large and compound, with 2–8 pairs of leaflets. Inflorescences are racemose, drooping, axillary, many flowered and 10–40cm long. Flowers bright yellow with 10 stamens. Three stamens very long, distinctly longer than other seven and curved up. When blooming, whole plant is covered by the inflorescences, and is very attractive and noticeable at a long distance. Fruit is a pendulous, black, indehiscent (non-splitting) legume to about 20–60cm long.

DISTRIBUTION Naturalized exotic species. Low and mid-elevations. Very common in dry zone, where it has been extensively planted as a roadside tree. **HABITAT** Low-country dry-zone forests. **FLOWERING PERIOD** July–August.

Pink Coral Shower ■ *Cassia grandis*

IDENTIFICATION Large, branched deciduous tree to about 15m tall. Leaves compound and long (about 25cm). Each leaf consists of 8–20 pairs of leaflets that are prone to being shed easily (caducous). Inflorescences are racemose, 10–20cm long and appear when tree is leafless. When flowering, whole tree is crowded by pink to light orange flowers, which usually coincide with the period when leaves are shed in this tree. **DISTRIBUTION** Low-country dry zone. Very common at archeological sites such as Polonnaruwa and Anuradhapura, where it has been planted for both shade and colour. **HABITAT** Cultivated widely as an ornamental plant. Common in dry and intermediate zones. **FLOWERING PERIOD** Generally, throughout the year, but more flowers apparent after the north-east monsoon.

Woolly Cassia ▪ *Cassia hirsuta*

IDENTIFICATION Almost all parts of this herb are covered with coarse and rather stiff hairs (hirsute), so its scientific name '*hirsuta*' is rather apt. Leaves compound, usually with 3–5 pairs of leaflets. Like the other Cassia species, this one produces terminal racemes with a few flowers. Flowers yellow, and 10 stamens per flower, two of which are exceptionally large compared with the others. Anthers brown. Fruit is a linear, curved, densely hirsute pod. **DISTRIBUTION** Exotic. Throughout Sri Lanka. **HABITAT** Moist places with good sunlight, for example along tributaries, paddy fields and roadsides. Also open land. **FLOWERING PERIOD** Photographed in October.

Feather-leaved Cassia ▪ *Cassia mimosoides*
(S: Bin-siyambala)

IDENTIFICATION Annual herb with an erect stem, which grows to about 1m tall. Leaves compound, about 10cm long. Each has 30–50 pairs of stalkless (sessile) leaflets. When touched, leaflets show slight movements and 'go to sleep'. Margins of leaflets, pedicel and underneath of sepals dark brown to scarlet. Flowers solitary in leaf axils. Petals and stamens yellow. Stamens (10 in number) nearly equal in length. Fruit is a flat, small pod with reddish to dark brown margins. **DISTRIBUTION** Native. Mid-country and low-elevation areas. **HABITAT** Moist or semi-dry open land. **FLOWERING PERIOD** Photographed in March.

Coffee-weed ▪ *Cassia occidentalis*
(S: Peni-tora; T: Ponnantakarai)

IDENTIFICATION Herb that grows to about 2m tall. Young stems, fronds, veins and margins reddish-brown. Compound leaf carries about six pairs of leaflets. When flowering, a few flowers are clustered in an inflorescence that arises from a leaf axil. Flowers bright

yellow, each with five petals and 10 stamens. Two stamens very large compared to others. Stamens crowded at apices of inflorescence and tinged purple. Fruit is a linear to slightly curved pod, tinged with purple. **DISTRIBUTION** Native. Dry lowlands. **HABITAT** Areas with good sunlight such as roadsides, tank bunds and abandoned sites. **FLOWERING PERIOD** January–February. Flowering usually seen after north-east monsoon.

Sickle Pod ▪ *Cassia tora*
(S: Peti-tora; T: Vaddutakarai)

IDENTIFICATION Annual herb that grows to about 1m tall. Young stems, rachis and flower stalks (pedicels) slightly hairy. Leaves compound, with 2–4 pairs of leaflets.

Terminal leaflet on tip of rachis larger than other leaflets. Flowers produced in small racemes in leaf axils. Pedicel about 1cm long. Flowers yellow, with five petals and 10 stamens. Five green sepals. **DISTRIBUTION** Native. Throughout Sri Lanka. **HABITAT** Open and disturbed places. **FLOWERING PERIOD** Photographed in March. Data on flowering not available.

Butterfly Pea ■ *Clitoria ternatea*
(S: Nil-katarodu, Katarodu-wel; T: Karuttappu)

IDENTIFICATION Slender vine that grows about 2m long. Leaves pinnately compound with five leaflets. Typically only one flower in an inflorescence. Two bracts at base of pedicel. Sepals form calyx tube. Petals of flower reminiscent of wings of a butterfly, and plant is also referred to as the butterfly plant. Its most striking feature is the colour of its flowers – vivid deep blue to light violet. In some cultivated forms there can be light yellow markings, which may vary from white to dark shades of blue and purple, and some varieties have double flowers. Fruit is a sparsely hairy, flat leguminous pod. **DISTRIBUTION** Native. Throughout Sri Lanka except hill country. **HABITAT** Moist, neutral soil in dry regions. **FLOWERING PERIOD** Throughout the year.

Crotalaria anagyroides
(S: Adanahiriya)

IDENTIFICATION Shrubby plant that grows to 1–3m tall. Leaves trifoliate compound. Racemes 15–30cm long and produced at tip of each branchlet. Fresh flowers bright yellow, old ones faded pinkish-yellow. Largest petal of flower is known as the standard,

and is ovate, bright yellow with blackish marks, later turning reddish at centre. Fruiting pod, borne on long stalk, 4–5cm long, cylindrical and pubescent. When fruit is shaken a crying sound can be heard – a feature common to all *Crotalaria* species. **DISTRIBUTION** Exotic. Throughout Sri Lanka except arid zone. **HABITAT** Disturbed open land such as tea plantations and roadsides. **FLOWERING PERIOD** Photographed in March.

Bird-flowers ■ *Crotalaria laburnifolia*
(S: Yak-beriya)

IDENTIFICATION Perennial herb that grows to about 2m tall. Erect stems. Leaves trifoliate with long petiole. Inflorescence is a terminal raceme with many flowers. In

the flower, standard is positioned upwards and keel is positioned downwards. Keel dark reddish with prominent beak. All other petals bright yellow. Pod

smooth (glabrous), about 2–3cm long, with 10–30 seeds. **DISTRIBUTION** Native. Low country and intermediate zone. **HABITAT** Open ground with good sunlight. **FLOWERING PERIOD** February–May.

Small Rattle Box ■ *Crotalaria pallida*

IDENTIFICATION Herb that grows to about 1m tall. Erect stems. Leaves trifoliate compound. Leaflets elliptic to obovate in shape. Inflorescences terminal or axillary, racemes with 20–30 flowers. Flowers (1.3cm long) yellow, often with reddish

or brown striations, especially on keel. Pods oblong-cylindrical and covered by tiny hairs, with 25–30 olive-green to brown seeds. **DISTRIBUTION** Native. Wet and dry zones. **HABITAT** Waste ground, along roads. **FLOWERING PERIOD** Photographed in March.

Rattle Weed ■ *Crotalaria retusa*
(S: Kaha-andana-hiriya; T: Kilukiluppai)

IDENTIFICATION Grows as either an annual or perennial herb, to about 1.3m tall. Stems erect, or sometimes lying flat on the ground (prostrate). The presence of a simple leaf is a special characteristic of this *Crotalaria* species over the other *Crotalaria* species. Leaf tip slightly lobed (sometimes an obtuse tip may appear). Inflorescences are racemes with many flowers. Flower about 1.5–2.5cm long, bright yellow, standard with reddish veins. Keel not very prominent. Fruit cylindrical, 3–5cm long, with 15–20 seeds. Persistent calyx in fruit tinged purple, with join of two carpels in fruit also purple. **DISTRIBUTION** Native. Low-country dry zone. Also coastal belt. **HABITAT** Wasteland, seashore. **FLOWERING PERIOD** Photographed in March.

Blue Rattlepod ■ *Crotalaria verrucosa*
(S: Nil-andana-hiriya, Yakbairiya; T: Kdukiluppai)

IDENTIFICATION Annual herb that grows to about 1m tall. Stem much branched, erect or spreading, quadrangular shaped and hairy (pubescent). Leaves simple, rhomboid, with tiny hairs on both surfaces. Leaf margin undulate. Like other *Crotalaria*, this species has a terminal raceme as an inflorescence. Unlike in other *Crotalaria*, flowers are purplish to blue, and sepals purplish-green. Flowers of all other *Crotalaria* species are yellow with some other minor colour variations. Fruit oblong and cylindrical, 2–5cm long. Seeds yellowish to brown. **DISTRIBUTION** Native. Throughout Sri Lanka. **HABITAT** Moist locations in lower montane, wet- and dry-zone forests. **FLOWERING PERIOD** Photographed in January.

Rattlepod ■ *Crotalaria walkeri*

IDENTIFICATION Woody herb that grows to about 2.5–3m tall. Stems covered with minute white hairs. Leaves simple, elliptic to ovate, tip obtuse most of the time. Inflorescences are terminal racemes with a few flowers. Flower 1.5–2cm long. Sepals brownish-green – sepals that grow towards the ground are much longer than the others. Petals bright yellow with red venation. Entire midvein of standard petal reddish-brown. Keel of old flowers tinged pink. Fruit is sessile, about 4–5cm long. **DISTRIBUTION** Native. Upper elevations of montane forests. **HABITAT** Common in cold, wet, misty areas in upper montane forests. **FLOWERING PERIOD** Throughout the year.

Rattlepod ■ *Crotalaria wightiana*

IDENTIFICATION Shrubby herb that grows to about 1m tall. An identification feature of this species over other *Crotalaria* species is the dense, shining rusty or golden hairs that

are nearly flat against (subappressed) over the stems, leaves, bracts and sepals. Leaves simple, elliptic-ovate, usually obtuse, rounded at base, and to about 5–6cm long. Inflorescences are lateral, racemes with 2–5 flowers. Sepals brown. Petals yellow. **DISTRIBUTION** Native. Higher elevations in montane and intermediate-zone forests. **HABITAT** Patana and open scrubland. **FLOWERING PERIOD** March, August.

Zanzibar Rattlepod ■ *Crotalaria zanzibarica*

IDENTIFICATION Grows as large annual or perennial herb, to about 3m tall. Stems erect and ribbed. Inflorescence is a terminal raceme with many crowded flowers. Petals orange or bright yellow, with purple stripes or dark spots near base. Fruit oblong-cylindrical, 3–4.5cm long, with many seeds. **DISTRIBUTION** Exotic. Higher elevations of intermediate zone and lower elevations of montane zone. **HABITAT** Common on waste ground and roadsides; hill country. **FLOWERING PERIOD** March.

Derris parviflora
(S: Kala-wel, Sudu Kala-wel)

IDENTIFICATION Grows as a liana. Leaves compound, with 5, 7 or 9 leaflets, which are coriaceous. Inflorescence very complex and like a panicle. Flowers about 5–7mm long and

sweet scented. Calyx reddish and fringed with very small hairs (ciliolate). Petals white or pale purplish. Pod is sessile, one or two seeded, about 3–6cm long and narrowly winged along both margins. **DISTRIBUTION** Endemic. Lowland dry-zone and intermediate-zone forests. **HABITAT** Favours dry locations with good sunlight. **FLOWERING PERIOD** May, September.

Spanish Clover ■ *Desmodium heterophyllum*
(S: Maha-udupiyali)

IDENTIFICATION Ground-dwelling herb that spreads through underground stolons. Young branches, pedicel and leaf margin crowded with white shining hairs. Leaves trifoliate. Leaflets thin and stiff, like paper (chartaceous), elliptic and slightly lobed at apex with a notch (emarginated).

Only a few flowers at a time are in bloom on an inflorescence. Flower about 1cm in width. As in many legumes, flower has a standard, keel and two wings as petals. Petals purple with white base. Plant is famous as rabbit fodder. Also grown as a cover crop in home gardens. **DISTRIBUTION** Native. Throughout Sri Lanka. **HABITAT** Roadsides and open places, moist grassland. **FLOWERING PERIOD** June–October.

Sickle Bush ■ *Dichrostachys cinerea*
(S: Andara; T: Viddaththal)

IDENTIFICATION Small, tree-like shrub that grows to about 3m tall. Leaf bipinnate, compound, with 8–16 pairs of leaflets (pinnae). Leaflets turn when touched. Sharp thorns on branches. Flower

is a two-coloured, drooping raceme. Florets are just like threads. Florets near to pedicel pink, further (distal) florets yellow. **DISTRIBUTION** Native. Lowland dry-zone scrub forests and arid-zone scrub. **HABITAT** Very common on roadsides, scrubland and forests in lowland dry zone. **FLOWERING PERIOD** March.

Gal Karanda ■ *Humboldtia laurifolia*
(S: Gal Karanda, Ruan-karanda)

IDENTIFICATION Grows as a shrub. A special feature of the plant is the presence of swollen internodes that harbour ants. Leaves compound with 5–8 pairs of leaflets; no

terminal leaflet. Leaflets about 10cm long, dark green and have an acuminate tip. Numerous white flowers clustered in upright racemes. Petals white. Stamens project beyond corolla (exserted). Anthers reddish to brown in colour. **DISTRIBUTION** Native. Lowland rainforests. **HABITAT** Moist, wet-ground vegetation of lowland rainforests. **FLOWERING PERIOD** April.

Sensitive Plant ■ *Mimosa pudica*
(S: Nidikumba; T: Tata-vadi)

IDENTIFICATION Creeping herb that can be annual or perennial. Sparse thorns on stem. Compound leaves fold inwards and droop when touched, hence the name 'sleeping plant', or 'nidicumba', given to the plant. Flowers are pink heads and arise from leaf axils.

Flower heads nearly 2.5cm in diameter. Number of flowers per plant higher in older plants than younger ones. Ball of stamens radiates from flower and is its dominant feature. Root system has small, light brown nodules. **DISTRIBUTION** Exotic. Throughout Sri Lanka. **HABITAT** Very common weed of roadsides, wasteland and open ground. Most common in wet and intermediate zones. **FLOWERING PERIOD** Throughout the year.

Purple Tephrosia ■ *Tephrosia purpurea*
(S: Gam-pila,Pila; T: Kavali, Kolinchi)

IDENTIFICATION Herbaceous plant that has either an erect stem or uses a spreading stem like a runner. Sometimes stem base is woody. Leaves pinnate, compound, with 4–10 pairs of leaflets that are obtuse elongate and have a slight lobe at tip. Leaf tip ends with short, stiff point (mucronate). Flowers (nearly 1cm across) occur in a terminal raceme, and are purple-red to pink. Flower colour varies from location to location. Pod flat and small (3–5cm long). **DISTRIBUTION** Native. Widespread in lowland dry zone, scrubland and coastal areas. **HABITAT** Common plant found on roadsides and wasteland, and in coastal areas in dry zone. **FLOWERING PERIOD** February.

Gorse ■ *Ulex europaeus*

IDENTIFICATION Rapidly spreading shrub with spiny stems and branches. Spines are in fact reduced leaves, and are dark green, long and very sharp. Plant is so well defended by spines that it is left untouched by herbivores and can form impenetrable thickets. In sites such as Horton Plains National Park, where it is invasive, it has to be removed through interventionist conservation management. When flowering, plant adds a splash of colour to grassland, contributing to the scenic beauty of the highlands. Flowers (2cm long) bright yellow. Sepals pale yellow and hairy (pubescent). **DISTRIBUTION** Exotic. Upper hill country and upper montane forest areas like Horton Plains National Park, Nuwara Elya and Ambewela. **HABITAT** Invasive plant found in humid, wet and cold locations. Common, spreading along forest edges, roadsides and open forest patches. **FLOWERING PERIOD** Latter months of the year.

ROSACEAE (ROSES)
The Rosaceae is probably one of the best-known plant families in the world due to the cultivation of roses in gardens, and their visual appeal. The family includes trees, shrubs and herbs, and there are more than 3,000 species in around 100 genera. Most plants have the familiar five petals and numerous stamens. Although the individual flowers have the same basic model, the manner in which the flowers appear is variable, from individual flowers as in a garden rose, to flowering spikes and dense clusters of flower heads. To the casual eye they can look unrelated. The shapes of the fruits in different species are also variable.

Mock Strawberry ▪ *Duchesnea indica*

IDENTIFICATION Herbaceous perennial creeper that spreads rapidly with the help of stolons. Roots and crown of leaves arise from each node. Each crown has a few compound leaves. Leaves trifoliate, with long petioles. Margins of each leaflet have rounded, symmetrical teeth (crenate). One or more flower stalks arise from crown, and each stalk produces one flower. Tiny flower (2.5cm in diameter) has five green sepals and five attractive yellow petals. Stamens, anthers and all other parts of flower yellow. Each flower is replaced by a showy, bright red berry that is spheroid or ovoid in shape. Sepals remain on

berry. Ripened berry of wild strawberry is edible, but not as sweet as a cultivated strawberry. **DISTRIBUTION** Native. Nuwara Eliya, Ambewela, Horton Plains National Park, Hatton area and Peak Wilderness Sanctuary. **HABITAT** Undergrowth of montane forests. Most commonly seen in wet patana grassland. Also common on roadsides, and in tea plantations, vegetable farms and disturbed areas in the Central Highlands, where the climate is cold. **FLOWERING PERIOD** December–February.

Oval-leaved Bramble ▪ *Rubus ellipticus*
(S: Nara-bute)

IDENTIFICATION Scrambling shrub. Sometimes grows by using the support of nearby plants. Young stems covered in long, stiff hairs (hispid). Stems and branches of reddish bristles conspicuous for this species. Prickles scattered, strong, hooked and recurved. Leaves trifoliate, compound with elliptic leaflets. Terminal leaflet larger than other two leaflets. Inflorescence is a much-branched terminal panicle with many flowers, each about 1.2cm across. Five sepals, green and covered underneath by stiff white hairs. Calyx lobes are persistent on fruits. Five pure white petals slightly longer than sepals. Stamens and carpels clustered at centre of flower in

disk-like arrangement. Fruit round, about 1cm in diameter, green when immature and orange-yellow at ripening. **DISTRIBUTION** Native. Hill country. **HABITAT** Open places, roadsides, edges of jungles and disturbed habitats. **FLOWERING PERIOD** Throughout the year.

Mountain Bramble ▪ *Rubus fairholmianus*
(S: Katu buta)

IDENTIFICATION Scrambling shrub to about 3m long. Stem whitish, with covering of short hairs. Old stems are woody. Prickles sparse, distant, weak and slightly recurved. Leaves simple, 5–13cm long, with margin shallowly lobed into five or seven lobes. Reticulate venation very clear on leaf. Surface of leaf green and underneath pale, becoming dirty white with age. Inflorescence is a terminal panicle with many flowers. Five calyx lobes, 1cm long and triangular. Outside has short, thin hairs (pilose), and inside densely covered with short, soft hairs (tomentose). Calyx is persistent on fruit. Five white petals slightly longer than sepals. Many carpels and stamens clustered at centre of flower. Fruit nearly round (subglobose), to 1.3cm in diameter, and orange, turning glossy red when ripe. A few drupelets in

each fruit. **DISTRIBUTION** Native. Montane zone at about 2,000m. **HABITAT** Cleared forest patches, forest edges, roadsides and similar areas. **FLOWERING PERIOD** November–March.

White-fruit Bramble ■ *Rubus leucocarpus*
(S: Wal rosa)

IDENTIFICATION Like other *Rubus* species, this one is a scrambler. Young branches have long, soft hairs (villose), and numerous prickles that are strong, broad based and recurved. Leaves compound, pinnate. Five or seven pinnate leaflets with a single terminal leaflet (imparipinnate). Prickles also found in rachis and petiole. Secondary veins clear on leaflets. Flowers (1cm wide) produced in a terminal cyme. Five calyx lobes, ovate-triangular, tip with sharp point, velvety inside and outside, armed with a few straight

prickles. Five petals nearly equal in size to sepals, and deep pink. Numerous stamens. Anthers pink to purplish. Many carpels clustered at centre of flower. Fruit round (globose), 1–1.6 cm across, greenish-white (purplish-blue when ripe), fleshy and covered with thick, cottony white covering of downy hairs (tomentum). **DISTRIBUTION** Native. Upper montane zone. **HABITAT** Open thickets, roadsides and jungle patches. **FLOWERING PERIOD** December–March.

> ### CUCURBITACEAE (CUCUMBERS)
> The cucumbers are among the most important plant families for food, and include gourds, squashes and melons. There are about 800 species in 120 genera, with a cosmopolitan distribution. The plants have characteristic five-sided stems and coiled tendrils. Many species have leaves that are palmately lobed, and most are annual vines. The flowers are often large and showy. The only tree in the family is the cucumber tree that is endemic to Socotra, an island off the Horn of Africa.

Scarlet Gourd ▪ *Coccinia grandis*
(S: Kowakka; T: Kovval)

IDENTIFICATION Perennial plant that is dioecious – that is, the flowers are unisexual, with male and female plants producing single-sex flowers. Stem slender and herbaceous, becoming woody and papery with age. Plant grows aggressively and is able to carpet canopies and ground vegetation. It has tendrils, and the palmate leaves are variable between plants, shallowly or deeply lobed. Tip of each lobe can be pointed or rounded (obtuse). Leaf sometimes has secondary lobes. Flowers (5cm in diameter) solitary, or more rarely occur in pairs. Corolla pure white and throat light green. Corolla covered with dense white hairs. Flower has floral tube formed from the petals, as well as calyx tube. **DISTRIBUTION** Native. Wide distribution in lowlands, from sea level to 800m. **HABITAT** Woodland, forest borders, secondary forests and wasteland. Often seen growing on fences and utility poles. **FLOWERING PERIOD** June–September.

Spiny Gourd ■ *Momordica dioica*
(S: Thumba karawila)

IDENTIFICATION Perennial, dioecious plant – that is, the unisexual male and female flowers are produced on different male and female plants. Stem slender and much branched. Plant has tendrils and can climb by using the support of nearby plants.

Leaves simple with long, slender petiole. Leaf palmately lobed, with secondary leaf lobes. Yellow flowers (2–3cm in diameter) solitary in leaf axils. Male flower has large bract, which female flower lacks. Fruit small (3.5 x 2cm), oblong-ovoid, beaked, and covered overall with conical, soft, small protuberances (papillae). **DISTRIBUTION** Native. Dry areas from lowlands to about 1,500m. **HABITAT** Edges and forests in dry zone. **FLOWERING PERIOD** Photographed in March. Information on flowering in Sri Lanka absent in available literature.

Snake Gourd ■ *Trichosanthes cucumerina*
(S: Pathola)

IDENTIFICATION Annual, dioecious climber. Stems slender, with tendrils on each node. Leaves simple, nearly circular, hairy on both sides. Leaf base cordate and palmately

lobed. Star-shaped flowers 6cm across, white and very attractive, with spreading petals that have an extraordinary fringe. Cultivated plant, with fruit being consumed as a vegetable (snake gourd). **DISTRIBUTION** Cultivated. Low-country dry zone and mid-elevations in wet zone. **HABITAT** Found in cultivation. **FLOWERING PERIOD** Flowering period variable, depending on cultivation conditions.

COMBRETACEAE (COMBRETALES)

This mainly tropical family of lianas, shrubs and trees extends its range into the subtropics and warm temperate regions. There are about 500 species in around 20 genera, and they are either evergreen or deciduous. A characteristic of the family are the unique, unicellular, compartmented hairs on the leaves, which are alternate or simple, without stipules or with vestigial stipules. Flowers in inflorescences are typically radially symmetrical, and most species are bisexual, but some are unisexual. Certain species, such as the Indian Almond *Terminalia catappa*, have edible fruits. Many are economically important for their timber.

Sea-almond ■ *Terminalia catappa*
(S: Kottamba)

IDENTIFICATION Tall (10–35m), deciduous tree. Leaves very large (nearly 20cm long), thick, arranged spirally, and clustered and crowded at ends of branches. Shape of leaves usually obovate or elliptic-ovate. Gummy texture can be felt on surfaces of mature leaves. White, stalkless (sessile) flowers produced on axillary spikes, and may be male or female. Spike usually dominated by male flowers; a few female flowers can be seen at base of spike. Male and female flowers can be distinguished by looking for the stamens – male flowers have stamens, while female flowers lack them. Flowers tiny (0.5cm in diameter). Mature fruit large, ellipsoid or ovoid, laterally more or less compressed, glabrous and ringed by rigid wing. Seeds are edible. **DISTRIBUTION** Native. Coastal areas, inland and uplands to altitudes exceeding 800m. **HABITAT** Common in lower elevations and coastal areas. Also in mid-elevations and widespread in urban habitats. **FLOWERING PERIOD** Most months of the year, especially after the monsoons.

ONAGRACEAE (EVENING PRIMROSES OR WILLOWHERBS)
This family is found in warm temperate and tropical areas, and species diversity is highest in the New World. There are about 650 species in around 18 genera, growing as herbs, shrubs and trees. They occur in open habitats, and a few species are aquatic. The flowers usually have four sepals and four petals. The bisexual flowers are axillary in inflorescences (like spikes and racemes). Many plants in this family, such as fuchsias, are popular in gardens, and are native to South and Central America.

Peruvian Primrose-willow ■ *Ludwigia peruviana*
(S: Berudiyanilla)

IDENTIFICATION Shrubby plant that grows to 1m tall. Entire plant covered with long, soft hairs (villose), and every margin of plant appears white because of the hairs. Leaves lanceolate, 4–12cm long. Secondary veins parallel to each other. Most leaf margins are upcurved. Solitary flowers occur in leaf axils and have five sepals and petals. Petals bright yellow, with prominent veins, flat and nearly circular (suborbicular), 18–24mm long and shallowly notched (emarginated). Eight anthers on short filaments. **DISTRIBUTION** Exotic. Disturbed places at 150–1,500m. **HABITAT** Common in moist, open, muddy places, for example beside streams and in floodplains; also in paddy fields. **FLOWERING PERIOD** Throughout the year.

MYRTACEAE (MYRTLES, EUCALYPTUS, CLOVES & GUAVAS)
This large family comprises about 6,000 species in around 140 genera, and contains many familiar plants. It is mainly tropical, but extends its range to the temperate zone. Most of the species grow as tall trees or shrubs. The plants' leaves are typically opposite, but also sub-opposite and spirally arranged, and typically simple without stipules. They contain translucent gland dots with ethereal oils that when crushed are aromatic. The flowers occur in many forms, and can grow in inflorescences or as solitary flowers. They have 4–5 sepals and petals, and many showy stamens. The fruits may be dry or fleshy.

Cherry Guava ■ *Psidium cattleyanum*
(S: Embulpera)

IDENTIFICATION Tree that grows to about 4m tall. Bark brown and smooth. Leaves oval to elliptical, to 4.5cm long, and smooth and leathery to waxy. Flowers small (1cm across), white, with numerous anthers. Stigmatic surface greenish. Round fruits are depressed (flattened from above, downwards), fleshy, sweet and edible. Ripe fruit scarlet and shiny. Pulp white with many seeds. Sepals persist on fruits. **DISTRIBUTION** Exotic. Throughout Sri Lanka except extreme dry areas. **HABITAT** Secondary forests, home gardens and similar places. The plant is becoming invasive. **FLOWERING PERIOD** Early months of the year.

Sour Guava ▪ *Psidium guineense*
(S: Kelepera, Embulpera)

IDENTIFICATION Appearance that of a small tree, though sometimes grows more like a shrub. Leaves elliptic, deep green, velvety on both surfaces, about 14cm long, with clear veins. Leaf arrangement is opposite. Flowers 2cm across, pure white, fragrant and showy. Many stamens, each with a white anther and filament. Ovary is inferior (situated below petals and sepals). Fruit firm, rounded, to 2.5cm wide. Sepals persist on fruits. Skin of ripe fruits yellow and shiny, outer pulp yellow and whitish, and inner pulp contains many seeds. Fruit is edible. **DISTRIBUTION** Exotic. Wet-zone and low-hill country. **HABITAT** Common in secondary forests and widely cultivated. **FLOWERING PERIOD** Throughout the year.

Wild Guava ▪ *Rhodomyrtus tomentosa*
(S: Seethapera, Sudu-kotala)

IDENTIFICATION Much-branched shrub that can be easily identified by dense covering of short hairs (tomentose) on whole plant, including stems and leaves. Stems covered by tiny brown hairs, and leaves by tiny white to grey hairs. Leaf about 8cm long, elliptic, stiff and leathery (coriaceous), drying to dark purple above. Basally three nerved, with many secondary veins. Veins prominent underneath leaf. Flowers (about 3cm across) are axillary, solitary or in simple racemes. Corolla pink, with five petals. Numerous stamens. Filaments pink and anthers yellow. Style pink. Fruit is edible, about 1cm in diameter, round and crowned with a persistent calyx. Seeds embedded in a purple pulp. **DISTRIBUTION** Native. Hill country above 1,500m. **HABITAT** Roadsides, grassland and open ground in cool hill country. **FLOWERING PERIOD** Early months of the year.

Indian Blackberry ■ *Syzygium caryophyllatum*
(S: Dan)

IDENTIFICATION Small tree (or shrubby plant) that grows to about 2m tall. Branches are ascending and branchlets are straight. Leaves obovate (about 10cm long), dark green, with smooth surface. Intermarginal venation clear and 1mm within margin. Inflorescence is terminal or subterminal; very complex. Minute flowers. Calyx cup shaped and corolla white. Numerous stamens, about 4mm long, with white filaments. Fruit about 8mm in diameter, nearly round (subglobose) in shape, and characterized by an apical collar. Ripe fruits dark purple and succulent. Fruits are edible. **DISTRIBUTION** Native. Wet-zone lowlands and intermediate zone. **HABITAT** Open places in secondary forests, open sandy areas and home gardens, and grown alongside cultivated crops such as tea. **FLOWERING PERIOD** February–May.

MELASTOMATACEAE (MELASTOMES)

This pantropical family comprises about 5,000 species in around 170 genera, which occur as herbs, woody climbers, lianas, epiphytes, shrubs and trees. Two-thirds are found in the neotropics, and they grow across a wide elevational gradient, from lowlands to cloud forests. In Sri Lanka, small shrubs from this family are common wayside plants. The venation of the leaves is a characteristic of the family – the lateral leaf veins are parallel to the margins and converge at the apex. The leaves are opposite, and decussate (alternate opposite pairs are at right angles to each other), and several species display anisophylly (a pair of opposite leaves at a node with a pronounced difference in size or shape). The flowers are showy and often in shades of purple. *Clidemia hirta*, native to the Caribbean, and Central and South America, is an invasive plant that has spread around the world.

Koster's Curse ■ *Clidemia hirta*
(S: Katakalu bowitiya)

IDENTIFICATION Grows as a branching herb and sometimes as a shrub, to about 1.5m tall. Leaves elliptic with a pointed tip, basally five veined, with many distinct lateral

veins. Veins crowded with prominent hairs, and young stems also covered with hairs. Small white (1cm) flowers borne in branched bunches. Ovary is inferior (below sepals and petals), cup shaped and green. Five sepals linear and hairy. Five petals stiff and leathery (coriaceous). Anthers white and narrowly linear. Fruit cup shaped. Ripe fruit deep violet-black and juicy. Fruit is edible and sweet in taste, but not readily eaten by people. An exotic, invasive weed that is spreading rapidly. **DISTRIBUTION** Exotic. Wet lowlands to highlands. **HABITAT** Disturbed areas, roadsides, and moist places along streams and rivers. **FLOWERING PERIOD** Throughout the year.

Malabar Melastome
■ *Melastoma malabathricum*
(S: Mahabowitiya, Katakaloowa)

IDENTIFICATION Native *Melastoma*, with the largest leaves and flowers, growing as a branched shrub. All young stems covered with brownish scales. Leaves elliptic and large (about 20cm long), with minute scales on both surfaces, so leaf surfaces are very rough. Leaves basally five nerved, with many distinct lateral veins. A few flowers at a time bloom in a bunch. Flower about 8cm in diameter. Five petals pink to mauve and purple; dark pink nerves are clear on petals. Ten stamens of two types: anthers of five stamens are yellow, while the other five stamens have jointed filaments and anthers that are pink to mauve. Fruit cup shaped. When ripe, fruits split open (dehisce). Pulp black and juicy. Fruit pulp is edible, but when eaten, the mouth is stained black. **DISTRIBUTION** Native. Throughout Sri Lanka. Common in montane and wet-zone locations. **HABITAT** Disturbed places, roadsides, riverbanks and secondary forests. **FLOWERING PERIOD** Throughout the year.

Rough Small-leaved Spider Flower ■ *Osbeckia aspera*
(S: Bowitiya)

IDENTIFICATION Highly branched shrub that grows to about 2m tall. Young branches hairy. Leaves large in comparison to leaves of other *Osbeckia* species. Three nerves are clear in leaf. Flowers borne in axillary bunches. One flower in bloom can be seen in a bunch at a time. Showy, pink to mauve flowers. Five sepals with bristles between sepals. Anthers yellow and very narrow. **DISTRIBUTION** Native. Common in dry zone and lower montane locations. **HABITAT** Grassland, open forests, on rocks. **FLOWERING PERIOD** March.

Eight Stamen Osbeckia
■ *Osbeckia octandra*
(S: Heen bowitiya, Bowitiya)

IDENTIFICATION Branched shrub that grows to 1m tall. Leaves about 6cm long, elliptic, three veined. White, shiny bristles on leaf surface. Pink to mauve flowers about 5cm across. Can very easily be distinguished from other *Osbeckia* species by its twisted anthers. **DISTRIBUTION** Endemic. Common *Osbeckia* species that can be seen all over Sri Lanka, and is most prevalent at lower elevations. **HABITAT** Roadsides, open places and grassland. **FLOWERING PERIOD** Throughout the year.

Small-leaf Osbeckia ■ *Osbeckia parvifolia*
(S: Wel-bowitiya)

IDENTIFICATION Climber, using runners to spread. Running habit and four petals in

flower help to distinguish it from all other *Osbeckia* species. Stem brownish-purple, covered by brownish to yellowish hairs. Leaves elliptic, three nerved and densely covered by long hairs on both sides. A few flowers bloom at a time. Flowers pink to mauve, with petals and sepals in sets of four (tetramerous), but occasionally in sets of five (pentamerous). **DISTRIBUTION** Native. Montane-zone forests and grassland. **HABITAT** Open places and grassland. **FLOWERING PERIOD** April, June.

Ruddy Osbeckia ▪ *Osbeckia rubicunda*

IDENTIFICATION Branched shrub to 2m tall. Can be easily distinguished from other *Osbeckia* species by its brownish-red sepals and brownish-red bristles over the surface of an inferior ovary. Branchlets densely hairy with brown hairs. Leaves elliptic (about 6cm long), five nerved, densely hairy. Flowers 5–6cm across, pink to mauve. Anthers yellow-brown mixed, and slightly spiral. Five sepals triangular in shape and brownish-red. Surface of inferior ovary covered by brown-green bristles. **DISTRIBUTION** Native. Very common in montane-zone forests and grassland. **HABITAT** Roadsides, open places, grassland and edges of montane-zone forests. **FLOWERING PERIOD** November, December, March, April.

Sadaraja ▪ *Sonerila pumila*
(S: Sadaraja)

IDENTIFICATION Small herb, branched very close to the ground. Stem quadrangular in shape and reddish-brown. Leaves 3cm long, symmetric, base oblique or cordate. Leaf margin shallowly dentate and ciliate. Leaf surface scattered with coarse white hairs that appear as irregularly distributed hairs. Petiole brownish-red. Flowers solitary, or a few flowers in inflorescence. Petals pink, 4–5mm long. Anthers ovate, blunt, with two large elongated pores. Capsule urn shaped (urceolate), with persistent sepals as long as pedicel, smooth, faintly ribbed. **DISTRIBUTION** Endemic. Montane-zone forests. **HABITAT** Common in montane-zone forests in shady areas, and along secondary generating areas of roadsides. **FLOWERING PERIOD** Photographed in May.

SAPINDACEAE (SOAPBERRIES)
This family is found mainly in the tropics and southern latitudes, but is also taken to include temperate species previously treated as being part of the Aceraceae and Hippocastanaceae. The family is best known for its trees, but also includes woody climbers and shrubs. In the northern temperate zone, the species are mainly in the genera *Acer* and *Aesculus*. There are about 2,000 species in around 150 genera. India has around 37 species and the Malesian region has about 235. The plants' leaves are alternate or opposite, and occur in a variety of forms, from pinnate to palmate, with different numbers of lobes.

Balloon Vine ■ *Cardiospermum halicacabum*
(S: Penelawel)

IDENTIFICATION Climbing herb. Leaves compound, margins of leaflets highly lobed, leaf petiole narrowly winged. Inflorescences each have a pair of tendrils that are coiled inwards (circinate). Flowers white and 4mm in diameter. Capsules nearly round, spinning top-shaped (turbinate), 1–3cm long, and winged at angles. Seeds are black. **DISTRIBUTION** Native. Moist and dry regions of lower elevations. **HABITAT** Wasteland, home gardens, moist and dry regions. **FLOWERING PERIOD** Throughout the year.

RUTACEAE (CITRUSES)
This familiar family of shrubs and trees includes the lemon, orange and grapefruit, which are cultivated extensively for their fruits. Perhaps somewhat surprisingly, the familiar wood apple is also a citrus. The family contains about 900 species in 150 genera, which are found mainly in tropical and warm regions. Many of the plants have aromatic leaves that when crushed give rise to a 'citrus' fragrance. The family is especially diverse in Australasia.

Ankenda ▪ *Acronychia pedunculata*
(S: Ankenda)

IDENTIFICATION Small tree. Leaves simple with a long petiole of about 5cm. Terminal inflorescence about 30cm long. Flowers greenish-white, 1cm across. Four pubescent petals in a flower. Petals curved towards peduncle. Eight stamens per flower. Anthers white. Ovary prominent, pale or tawny. Fruit nearly rounded (subglobose) in shape, and creamy to tawny in colour. **DISTRIBUTION** Native. Sea level to 1,600m. **HABITAT** Common in moist locations. **FLOWERING PERIOD** February–April, June, July.

Woodapple ■ *Limonia acidissima*
(S: Diwul; T: Vila, Vilatti, Mayladikkuruntu)

IDENTIFICATION Tree that grows to about 7m tall. Bark pale grey or whitish. Spines on branches about 4cm long. Leaves pinnately compound, with single terminal leaflet (imparipinnate), 10cm long with winged rachis. Inflorescences arise from nodes. Flowers numerous and very small. Petals white or pale green. Stamens conspicuous, 10 per flower, relatively larger, and pinkish or brownish in colour. In perfect flowers, green ovary and style are very clear. Fruit woody, round and pale, with sticky pulp and many seeds. Fruit is edible, and is used to make a delicious fruit drink, and in jams, ice creams and other preparations. **DISTRIBUTION** Native. Forest plant in dry-zone forests. Also cultivated in many home gardens in dry zone. **HABITAT** Dry habitats such as forests and home gardens. **FLOWERING PERIOD** Photographed in February. Common and one of the best-known wild trees in Sri Lanka, but there is a surprising absence of information in the published literature on its flowering period (this illustrates how much more remains to be studied on Sri Lanka's plants).

Orange Jessamine ■ *Murraya paniculata*
(S: Etteryia)

IDENTIFICATION Small tree with pale bark. Leaves pinnately compound with 3–7 leaflets and single terminal leaflet (imparipinnate). There are a few flowers in terminal inflorescences. Five petals pure white. Ten stamens. Columnar style. Stigma enlarged and bilobed.

DISTRIBUTION Native. Moist and dry zones of low country to 1,000m. **HABITAT** Low-country dry and wet locations. **FLOWERING PERIOD** Photographed in October. No published data on flowering period.

Orange Climber ■ *Toddalia asiatica*
(S: Kudu-miris; T: Kandai)

IDENTIFICATION Small climber. Young
stems covered with dense brown hairs.
Sharp, recurved prickles on stem. Leaves
tripinnately compound, and prickly
underneath. Leaflets stalkless (sessile), and
each about 5cm long. Flowers tiny, each
about 3–4mm long. Sepals very small and
brownish-green. Petals larger than sepals,
with white upper sides. Five stamens.
Fruit nearly round (subglobose), 3–5
grooved, glandular and orange when ripe.
DISTRIBUTION Native. Montane forests.
HABITAT Disturbed and undisturbed sites
in montane forests. **FLOWERING PERIOD**
Photographed in April. No published data
on flowering period.

MELIACEAE (MAHOGANIES)

Found throughout the tropics and subtropics, the Meliaceae family comprises about 550 species in around 50 genera. The plants grow mainly as shrubs and trees, and rarely as herbaceous shrubs with a woody stock. The family contains some very important timber trees in the genera *Swietenia* and *Khaya*, which are widely cultivated for commercial timber. The plants bear leaves that are alternate or spiral, usually pinnate and rarely bipinnate. The radially symmetrical flowers are functionally unisexual. The inflorescences are cymose panicles and display considerable variation. They can be terminal, or appear on the trunk or branches, or in leaf axils. The sepals and petals may be united or distinct, and the fruits may be capsules, drupes, berries or (rarely) nuts. The Kohomba *Azadirachta indica*, known as the Neem Tree in India, has been used in Ayurvedic medicine for more than 2,000 years.

Bin Kohmba ▪ *Munronia pinnata*
(S: Bin Kohomba)

IDENTIFICATION Unbranched shrublet to 35cm tall, which can be recognized from afar by the scent of the flowers. Leaves form apical rosettes, and are compound with three leaflets. Middle leaflet longer than other two leaflets. Petiole long. Inflorescence is a thyrse (see Glossary, p. 20), 2.5–10cm long. Highly scented, pure white flowers (2–4cm across) open a few at a time. Each flower has petals and a long white floral tube (staminal tube) at its centre. Further (distal) ends of stamens are free. Anthers yellow to yellow-brown. Top of staminal tube has a yellowish-brown stigmatic surface. **DISTRIBUTION** Native. Forests in wet, intermediate and dry zones to 700m. **HABITAT** Indigenous plant that grows on the ground and on rocks. Has recently become rare because of over-exploitation for commercial purposes. **FLOWERING PERIOD** Photographed in August. Information on flowering is sparse.

THYMELAEACEAE (DAPHNES)
The daphnes are a cosmopolitan family of about 800 species in 45 genera, found in the tropics and temperate zones, with species diversity being greatest in Africa and Australia. The plants grow as herbs, lianas, shrubs and trees. Their bark contains long fibres. Their leaves lack stipules, and are simple and alternate. The phloem and long fibres give the leaves and stems a certain rigidity. The radially symmetric flowers can be unisexual, bisexual or hermaphroditic (containing both unisexual and bisexual flowers in the same plant), and can occur in racemes, follicles or heads that are densely packed with stalkless flowers. The phylogeny of this plant group is still in flux, with some authors recognizing the Thymelaoideae (found in India and Sri Lanka), Aquilarioideae, Gilgiodaphnoideae and Gonystyloideae as subfamilies. The family was historically placed in the Myrtales order, but is now in the Malvales.

Tie Bush ▪ *Wikstroemia indica*

IDENTIFICATION Shrubby plant that grows to 0.5–1m tall. Old stems and branchlets reddish-brown in colour. Leaves simple, leathery, luminous green, elliptic in shape and oppositely arranged along stem. Yellowish-green flowers borne in terminal clusters; many flowers can be seen at once. Each flower (0.5–1cm in diameter) has four petals that are fused at the base, forming a floral tube. Stamens are inside floral tube. Anthers appear orange because colour of mature pollen is orange. Tiny fruits (0.5–0.75cm long) ellipsoid, fleshy, scarlet and shiny. Fruits and seeds emit an unpleasant odour when crushed. Birds refrain from eating them, and leaves and fruits are poisonous. **DISTRIBUTION** Native. Nawalapitiya, Gampola, Kandy, Matale. **HABITAT** Exotic. Open areas such as roadsides and rocky ground with direct sunlight are the best habitats. Commonly seen along roadsides of mid-country tea plantations. **FLOWERING PERIOD** Latter part of the year.

BIXACEAE (ANNATTOO)
This neotropical family contains just one genus and five species, which grow as shrubs or small trees. The sap in the stems and leaves can be yellow, orange or reddish. *Bixa orellana* is widely cultivated in the tropics, and may not even be found in the wild where it originated. The leaves have long petioles that are entire and simple, and can fall off prematurely. Their veins are palmate. The flowers are radially symmetric, and appear showy in thyrsoid inflorescences. They each have five sepals and five petals. The fruit is a capsule that splits longitudinally along the dorsal sutures of the wall (loculicidal). The economic use of the plant principally involves the extraction of the reddish dye and condiment annattoo, which contains mainly bixin.

Silk-cotton Tree ▪ *Cochlospermum religiosum*
(S: Ela-imbul, Kinihiriya; T: Kongu)

IDENTIFICATION Deciduous tree. Leaves tripalmately or palmately lobed. Petiole very long. Bright yellow flowers produced in terminal branches as inflorescences. Each flower

has five petals, with petal tips slightly bilobed. Petals marked with minute red streaks on both surfaces. Filaments long and clustered at middle of flower. **DISTRIBUTION** Exotic. Recorded in several monsoon forest areas such as Sigiriya and Ritigala. Often planted in Buddhist temples. **HABITAT** Rocky places in dry and intermediate zone. **FLOWERING PERIOD** March–May.

> ### MALVACEAE (MALLOWS)
> As many as 1,500 species comprise this family, which includes the well-known garden flower the 'Shoe Flower', or Hibiscus. The term 'Hibiscus' also refers to one of the 75 genera in the family. Mallows occur in a variety of forms, including herbs, shrubs and trees. Their flowers have five petals and clusters of stamens on a single stalk.

Kapok Tree ■ *Ceiba pentandra*
(S: Pulun gas)

IDENTIFICATION Deciduous tree that grows to to 20m or even taller. Bole cylindrical and ash coloured. In large trees, buttresses are present. Leaves compound and palmately lobed, and 5–9 entire leaflets in a leaf. Leaf arrangement is alternate. During the dry period, the tree lacks leaves at the crown, and it normally flowers during this period. Flower clusters produced at end of each branchlet. Flowers (4cm in diameter) pendulous. Peduncle and sepals green. Sepals joined in a flower, making a cup tinged with yellow. Petals come out from this cup, and are white, tinged with pink. Five stamens with yellow anthers prominent in flower. Fruit (nearly 25cm long) is a type of capsule. Fruit green when young, becoming brownish as it matures, and narrow at both ends. Mature fruit normally half splits before it is shed, and white woolly kapok can be seen inside it. Seeds (5–7mm in diameter) black and round. The tree has an economical value, with kapok being widely used in many industries to make items such as pillows and mattresses. **DISTRIBUTION** Exotic. Humid locations to about 1,000m. **HABITAT** Prefers humid, dry locations. **FLOWERING PERIOD** Dry season.

Hibiscus furcatus
(S: Napiritta)

IDENTIFICATION Spiny herb that grows mainly like a shrubby runner. Stem brown, with scattered, slender, recurved prickles. Leaves palmately lobed. Veins prominent on

leaf underside, with prickles along them. Flowers solitary, showy and large (8–10cm in diameter). Brownish-green calyx and green epicalyx (see Glossary, p. 14). Corolla yellow with dark crimson centre. Capsule 1.2–1.5cm long, ovoid and pointed. Fruit surface crimson, and epicalyx persistent on fruit. Taste of fruit is acidic. **DISTRIBUTION** Native. Throughout Sri Lanka. **HABITAT** Secondary forests, forest edges, roadsides, and disturbed habitats with secondary generation. **FLOWERING PERIOD** March.

Mysore Fanpetals ■ *Sida mysorensis*
(S: Siriwedibevila, Sudubavila)

IDENTIFICATION Perennial herb with erect stems that grow to about 0.5–1.5m tall. Stems densely covered with hairs. Leaves simple, base cordate, apex acuminate. Veins

palmate. Flowers white and solitary, but sometimes orange to yellow-orange. Calyx has five sepals, triangular in shape and green. Corolla 10–12mm in diameter. Five petals slightly longer than sepals. Staminal tube in floral centre. Yellow anthers. **DISTRIBUTION** Native. Lowlands and mid-country to about 700m. **HABITAT** Waste places, along roadsides and in secondary vegetation. **FLOWERING PERIOD** March.

Tulip Tree ▪ *Thespesia populnea*
(S: Suriya; T: Kavarachu, Puvarachu)

IDENTIFICATION Medium-sized tree that grows to about 5m tall. Young twigs and petioles densely covered by brown or silver scales. Leaves large (20cm long), deep green, triangular (deltoid) or heart shaped. Leaf margin entire. Prominent light green veins on leaf surface. Petiole very long (10–20cm). Flowers axillary, solitary and large. Corolla with five petals. Petals yellow with scarlet centre. Staminal tube present with numerous stamens. **DISTRIBUTION** Native. Throughout Sri Lanka except highlands. Common in dry locations. **HABITAT** Cultivated as shade tree in many places. **FLOWERING PERIOD** April.

Burr Mallow ▪ *Urena sinuata*
(S: Heen epala, Patta epala)

IDENTIFICATION Strongly branched, woody herb growing to about 0.5–1m tall. Branches covered in downy hairs (pubescent). Leaves very variable, with most being more deeply divided than those in the Caesar Weed (p. 116). Extra-floral nectaries at leaf base. Flowers about 2.5cm in diameter. Petals pink with dark rose bases. Staminal column has mauve anthers. **DISTRIBUTION** Native. Throughout Sri Lanka and very common in dry zone. **HABITAT** Sandy seashores, roadside thickets, grassland surrounding secondary forest, open forests and degraded woodland in sandy areas. **FLOWERING PERIOD** Photographed in March.

Caesar Weed ▪ *Urena lobata*
(S: Patta epala)

IDENTIFICATION Branched herb growing to 0.5–1m tall. Mature stems are woody. Stems, petiole and leaves white to yellowish: pubescent. Leaves simple, slightly lobed, deep green above and greyish beneath. Leaves variable in size: those near the ground are larger than upper ones. Extra-floral nectaries at base of midrib beneath and occasionally at adjacent veins. Flowers solitary, 2–3cm in diameter, pink or rose-purple. Staminal tube has violet anthers. Fruit covered by stiff, hooked spines. When touched, fruits attach to clothes, demonstrating how they use animals to disperse. **DISTRIBUTION** Native. Throughout Sri Lanka except highlands. **HABITAT** Roadsides, waste places and areas of secondary growth. **FLOWERING PERIOD** Throughout the year, with high abundance in early months of the year.

CAPPARACEAE (CAPERS)
Also referred to as the Capparidaceae, this family comprises about 650 species in around 17 genera of herbs, shrubs, lianas and trees from warm or arid areas. Mustard oils are produced from some species. Small trees from this family are among the most common trees to be seen in popular national parks such as Yala and Wilpattu. Their leaves are alternate, and one genus, *Apophyllum*, has no leaves and has photosynthetic stems. The leaves are typically simple, and the inflorescences are variable, including racemes, panicles, corymbs and single flowers. The genus *Capparis*, found in Sri Lanka, includes species that are monoecious and dioecious.

Spreading Caper ▪ *Capparis divaricata*
(S: Torikei)

IDENTIFICATION Highly branching, spreading shrub nearly 5m tall. Small, straight green thorns in each node. Leaves simple, nearly 5cm long, entire and ovate shaped. Leaf arrangement is spiral. Flowers (4.5cm in diameter) solitary, showy and produced on branch axis. Petals white to cream coloured, with yellow bases. Petal base brownish-red or maroon in old flowers. Showy stamens, with a large number (up to 50) of them in a flower, and they are longer than the petals. Anthers violet. Fruit ellipsoid. Young fruit green, with colour changing to red with maturity. **DISTRIBUTION** Native. Semi-arid and arid forests. **HABITAT** Roadsides of dry and arid/semi-arid areas. Also open areas in dry lower elevations. **FLOWERING PERIOD** Early months of the year.

Ceylon Caper ■ *Capparis zeylanica*

IDENTIFICATION Highly branching shrub that grows to 10m tall. Branches are straggling. As in other *Capparis*, there are thorns near the nodes. Leaves simple, leathery, with entire surface shiny, and sometimes 10cm long. Leaf tip pointed. Flowers (about 10cm in diameter) solitary or paired in axillary branches. Upper pair of petals yellow at base, becoming reddish-violet with age. All other areas of flower are white. Numerous filaments have purple anthers. Flowers highly scented. **DISTRIBUTION** Native. Lower elevations and dry-zone forests. **HABITAT** Well-drained places and roadsides in dry forests or submontane/lowland forests. Often seen on game drives in dry-zone national parks. **FLOWERING PERIOD** Early months of the year.

Garlic Pear Tree ■ *Crateva adansonii*

IDENTIFICATION Deciduous tree that grows to 15–20m tall. Leaf compound with long petiole that has two glands above tip. Inflorescences bearing 17–30 flowers produced on small twigs. Floral stalk is as long as 4cm. Flowers white or cream, with four petals, and 18–22 or more stamens up to 5cm long. Filaments pale purple and anthers brown. Parts of the plant are salty. **DISTRIBUTION** Native. Low-altitude dry woodland and scrub. **HABITAT** Dry and arid locations. Familiar tree in dry-zone national parks. **FLOWERING PERIOD** March, April.

> ### DROSERACEAE (SUNDEWS)
> The sundews are a family of carnivorous plants that use sticky hairs in their leaves
> to trap their prey of small insects. The hairs secrete enzymes to enable the plants to
> digest this food. This is believed to be an adaptation to allow the plants to cope with
> the nitrogen-poor condition of the damp soils in which they grow. There are about
> 180 sundew species in three genera, found across a wide range of tropical, subtropical
> and temperate zone areas. They occupy aquatic to seasonally dry habitats. Their leaves
> are spirally arranged, forming a basal rosette in the widespread genus *Drosera*, and
> their radially symmetric flowers are hermaphrodite. The well-known Venus Flytrap
> *Dionaea muscipula* is confined to Carolina in the USA.

Tropical Sundew ■ *Drosera burmannii*
(S: Kadullessa)

IDENTIFICATION Fascinating carnivorous plant found in nutrient-poor sites. Grows
close to the ground. The plant itself looks like a flower, with the stem not being clear. Its
radial leaves form a closed rosette appressed to the soil, 0.8–3.5cm across; leaf spathulate.
Upper surface carries sticky (viscid) glands or tendrils, which are longest at the margins
and short at the middle of the leaf. The sticky tendrils can move fast and enable the plant
to capture and digest small insects. Leaves together with glands are crimson. Inflorescence
erect and nearly 3–20cm long. White flowers. **DISTRIBUTION** Native. Throughout Sri
Lanka. **HABITAT** Moist and sandy soil in montane and mid-elevation patana grassland.
FLOWERING PERIOD February, July.

Sundew ■ *Drosera indica*

IDENTIFICATION Green plant with weak, erect stem covered with glandular hairs. Leaves narrowly linear, hairy and about 1.5cm long. Leaf broader than petiole. Leaves of

this species are very different from leaves of other species, looking like stems. Young leaves have venation that is coiled inwards (circinate). Old leaves near base of stem provide support to weak stems by acting as prop-roots. Flowers mauve or pinky-purple, usually over 1.5cm across and on axillary peduncles that are 3–15cm long. **DISTRIBUTION** Native. Wet lowlands, lower hill country and hill-country grassland. **HABITAT** Wet or intermediate patana grassland. Moist places especially near rice fields, swamps, shallow pools and ditches. **FLOWERING PERIOD** Information not available.

Shield Sundew ■ *Drosera peltata*

IDENTIFICATION Insectivorous plant with erect stem about 15cm tall. Leaves alternately spaced along whole length of stem, with a very slender, curved, nearly hair-like

(subcapillary) petiole 1.0–2.5cm long. Leaf blade semicircular with petiole attached in centre (peltate). Leaves carry on their upper (adaxial) surface numerous long, sticky glands, which are supported on special stalks (stipitate). Stipitate glands on leaf margin long and spreading, while the rest are short. White flowers nearly 1.5cm across. **DISTRIBUTION** Native. Upper montane zone. Grassland in mid-elevations. **HABITAT** Wet grassy patches on hill slopes and banks bordering waterways. **FLOWERING PERIOD** Information not available.

NEPENTHACEAE (PITCHER PLANTS)

This family of perennial tropical plants, growing as climbers, scramblers or epiphytes, is well known because of its carnivorous habits. Pitchers with a hood form on the ends of tendrils that are a continuation of the midribs on the leaves. Insects fall into the pitcher, attracted by sweet secretions, and cannot climb out of the slippery walls. The pitchers secrete digestive enzymes. Pitcher plants grow in nutrient-poor soils and can be seen in lowland rainforests, especially by the edges of roads and other clearings where nutrients have been leached away by rain. The male and female flowers grow in spike-like inflorescences, and are borne on separate plants (dioecious). The sepals and petals are fused into tepals, and the filaments of the stamens are fused into a column. The leaves are alternate and lack stipules. Young plants form a basal rosette from which a number of climbing stems may originate, and the tendrils are used for climbing. Pitcher plants have a centre of diversity in Borneo, and are also found in northern Australia, Southeast Asia, northern India, Madagascar and the Seychelles.

Pitcher plant ▪ *Nepenthes distillatoria*
(S: Bandura-wel)

IDENTIFICATION This well-known carnivorous plant can be easily identified from the pitcher at the tip of a tendril that extends from the leaf tip. The plant is a climbing liana. Leaves long (about 30cm). Leaf midrib extends and creates a cylindrical pitcher with a lid. Liquids and trapped insects can be seen inside a pitcher. Inflorescence is a panicle about 40cm long. Male and female flowers grow on separate plants. **DISTRIBUTION** Native. Wet lowland forests such as Sinharaja, Kanneliya and similar places. **HABITAT** Endemic plant common in primary and secondary forests, and scrubland. Found in nutrient-poor soils, especially in clearings where rain has leached the soil. **FLOWERING PERIOD** June.

THE NAMING OF A SRI LANKAN PITCHER PLANT

A pitcher plant common in Sinharaja was the second tropical pitcher plant species to be known to science. Jacob Breyn, a Dutch merchant, described it as *Bandura zingalensium*, drawing on its local name *bandura*. In 1683 the Swedish doctor Grimm described it as the *Plant mirabilis distillatoria*, or the miraculous distilling plant. Linnaeus in 1737 ascribed it to the genus *Nepenthes*, drawing on a passage in Homer's *The Odyssey* describing how Helen spiked the wine of her guests with the drug 'nepenthe'. Today this pitcher plant is known as *Nepenthes distillatoria*.

POLYGONACEAE (SORRELS)
This diverse family contains about 1,150 species in 50 genera, which occur mainly in temperate regions. The plants grow in many forms, such as herbs, shrubs, trees and lianas. Most species have flowers that are radially symmetrical. Their hard fruits are like seeds. The family includes the edible rhubarb as well as the Japanese Knotweed *Fallopia japonica*, which has become notorious in Europe and America as an invasive plant, originally from Asia.

Snake Thumb ■ *Persicaria nepalensis*

IDENTIFICATION Creeping or erect herb. Brownish-green or reddish-green angular

markings on leaf surface help to easily identify this plant. Inflorescence is a densely packed head of stalkless flowers (capitulum). Round pink capitulum is characteristic of the species and further helps with identification. Leaves simple and tapering at both ends. Petiole winged. Flowers tiny and pink. **DISTRIBUTION** Native. Wet montane areas. **HABITAT** Wasteland, cultivated land, roadsides of tea plantations and home gardens. **FLOWERING PERIOD** Throughout the year.

> ## PLUMBAGINACEAE (LEADWORTS, SEA LAVENDERS & THRIFTS)
> This is a widespread family of herbs, climbers and shrubs that grow as annuals and
> perennials. About 850 species in around 25 genera are found throughout the world,
> except Antarctica, and many are adapted to salty or sandy conditions. Their leaves
> are typically spirally arranged, often in basal rosettes, and in some species they have
> auricles (ear-like projections at the base). Many species have secretory glands on the
> leaves for getting rid of water, calcium, salts and mucilage. The arrangements of the
> radially symmetrical flowers can be diverse, with inflorescences being spicate, cymose,
> racemose or in dense clusters. The five sepals are fused to form a five-toothed or five-
> lobed tube. There are five free petals and five stamens. The plants yield extracts that
> are used in medicine, and are also cultivated in gardens.

Ceylon Leadwort ▪ *Plumbago zeylanica*
(S: Elanetul)

IDENTIFICATION Straggling shrub that is easily identified by the large-stalked, round,
crimson-tipped green glands on the green sepals. Leaves simple. Flowers pure white and
borne on a spike 30cm long. Corolla tube 2.2cm long. Petals obovate with a stiff, short
point (mucronate tip). Midvein of petals very clear. Stamens as long as floral tube. Anthers
projected (exserted) just beyond throat of floral tube. **DISTRIBUTION** Native. Low and
mid-elevation areas of dry and intermediate zones. **HABITAT** Common in dry locations of
low-country and mid-elevation areas. Also cultivated as an ornamental plant in many places.
FLOWERING PERIOD December–March.

NYCTAGINACEAE (BOUGAINVILLEAS)
Plants in this family range from the tropics to warm temperate zones, with the highest species diversity occurring in the Americas. There are about 400 species in around 30 genera, including many growth forms, such as herbs, climbers, shrubs and trees. Bougainvilleas are well-known woody climbers that have been introduced throughout the tropics for growing on the tops of walls and hedges. The simple leaves lack stipules, and are opposite, alternate or whorled. The flowers are unisexual or bisexual, and the plants bear showy inflorescences in panicles or cymes. The stamens are within a tubular perianth, with no petals. In the genus *Bougainvillea*, what look like three decorative and colourful sepals are in fact bracts. This genus of around 18 species is native to South America.

False Jalap ■ *Mirabilis jalapa*
(S: Sendrikka, Hendirikka)

IDENTIFICATION Branched herbaceous plant that grows to about 1m tall. Stems fleshy and swollen at nodes. Each leaf pair similar in size. Leaf pairs near base of plant are larger, while those near apex are smaller. Leaves triangular

with long petioles. Flower bunches produced at tips of branches. Flowers open in the evening and one flower blooms at a time. Each flower (3cm across) has five fused sepals, a floral tube and five fused petals. Anthers extend from floral tube. Flowers purple, pink, yellow or white. Mature fruit 5–10mm long, black, ribbed, and covered with small protuberances (tuberculate) between ribs. Sepals are persistent in fruit. **DISTRIBUTION** Native. Throughout Sri Lanka. **HABITAT** Wasteland, and cultivated in home gardens, roadsides and similar areas. **FLOWERING PERIOD** Throughout the year.

> ## Portulacaceae (Purslanes)
> The purslanes are a varied family of plants occurring as herbs, shrubs and trees.
> Around 400 species are found in tropical and other warm regions. The typically
> 4–6-petalled flowers have radial symmetry. Many of the plants are succulent, and some
> are edible. They are popular in horticulture.

Jewels of Opar ▪ *Talinum paniculatum*
(S: Gas-niviti, Rata niviti)

IDENTIFICATION Erect, shrub-like herb that grows to about 30cm tall. Leaves simple,
entire, stiff and leathery, spoon shaped and spirally arranged on stem. A few showy,
pale pink flowers bloom at a time in an
inflorescence. Five petals clearly separated from
each other, and each petal has a pointed tip.
Stamens clustered at centre of flower. Style split
into three (trifid) and pink. Fruit is round (3mm
in diameter) and pink. Leaves of this plant are
consumed as a vegetable. **DISTRIBUTION**
Exotic. Throughout Sri Lanka but most
common in dry and intermediate zones; absent
at high elevations. **HABITAT** Exotic plant
found in waste places, roadsides, bunds of paddy
fields and other shady areas. **FLOWERING
PERIOD** December–May.

Marsh Henna ▪ *Hydrocera triflora*
(S: Diya-kudalu, Wal kudalu)

IDENTIFICATION Perennial herb with an erect, succulent stem that grows to about 1m tall. Stem light green, sometimes with pink patches. Fibrous roots can be seen near plant base. Leaves simple and spirally arranged. Upper surface of leaves green, underneath pale green.

Leaves look stalkless. Two light green dots on both sides of leaf base. Leaves long, sometimes growing to nearly 30cm, and leaf margin toothed (serrate). Inflorescence is a raceme, containing 3–5 flowers. Flowers (6cm in diameter) bright pink tinged with reddish-purple. Deep interior of flowers is green. Inflorescence axis and peduncle pink. Fruit is a fleshy pseudo berry, pentagonal in cross-section, with rounded angles, short-beaked. Fruit pale green, becoming reddish or purplish-red at maturity. **DISTRIBUTION** Native. Marshland in dry-zone locations. **HABITAT** Prefers semi-aquatic habitats. Common in marshy places, stagnant pools and similar places. **FLOWERING PERIOD** March, April.

Stemless Balsam ▪ *Impatiens acaulis*

IDENTIFICATION Perennial herb with tuber-like small stem growing on the ground. A few leaves arise from stem. Leaves green, flat and circular (orbicular) in shape. Leaf base

cordate. Tiny hairs on both surfaces of leaf. When flowering, plant produces a raceme-type inflorescence that bears a few flowers at a time. Stalk of raceme and peduncles of flowers are brown. Flowers (3cm in diameter) pink or mauve with white centre. Five petals per flower, four of which are in the same plane, while fifth petal is folded into an upright position in contrast to the other four. Thread-like pink extension at back of flower. **DISTRIBUTION** Native. Hill-country forests at 750–2,000m. **HABITAT** Favours watery rocks in cold places in hill-country forests. **FLOWERING PERIOD** June, July, August.

Henslow's Balsam ■ *Impatiens henslowiana*

IDENTIFICATION Perennial large herb about 50–150cm tall. Stem densely branched. Leaves simple and lanceolate. Alternate, opposite pairs of leaves are at right angles to each other (opposite decussate). Leaf margins toothed (serrate). Petioles reddish. Tiny purple patches on leaf margin. Leaf scars clear on stem. Flowers solitary (5.5cm in diameter), often white with a slight pink or mauve tinge. Flower has five petals, and one of its sepals is elongated and forms a spur 4–8cm long. Young capsules green, mature ones brown. Mature capsules split and spread seeds. Most parts of the plant,

including young stems, leaves, petioles, peduncles and sepals, covered by short, thin hairs (pilose). **DISTRIBUTION** Native. Hill-country rainforests. **HABITAT** Wet rocky places or waterfall sprays. **FLOWERING PERIOD** Latter months of the year.

Ceylon Balsam ■ *Impatiens repens*
(S: Gal demata)

IDENTIFICATION Rare creeping herb that is endemic to Sri Lanka. Small, rounded, brownish-green leaves borne on red stems that are succulent, heavily branched out and create a thick mat of foliage. Flowers bright yellow and nearly 3cm in diameter. They have red markings inside, and interior end of a flower is curved, forming a small tail. **DISTRIBUTION** Endemic. Can be seen in the Royal Botanic Gardens at Peradeniya and in many home gardens. A few locations in the country where it occurs in

the wild have recently been discovered. **HABITAT** Wet places such as ponds and bogs with clean water. Natural habitat of this species was the wet-zone rainforests of Sri Lanka, but it has been very nearly lost in the wild. It has recently been cultivated in gardens as an ornamental plant. **FLOWERING PERIOD** Throughout the year.

> **LECYTHIDACEAE (BRAZIL NUTS)**
> This family comprises shrubs and trees, with about 300 species in around 25 genera.
> They are mainly found in the neotropics, with some genera also occurring in Africa,
> Asia and Australia. The leaves of the plants are spirally arranged or are in two ranks.
> They are simple, large and usually without gland dots. The inflorescences vary from
> racemes on the leaf axils, to panicles or long spikes. The Cannonball Tree *Couroupita
> guianensis* has a flowering spike that is 1m long. The family includes the well-known
> Brazil Nut *Bertholletia excelsa*.

Fish Poison Tree ▪ *Barringtonia asiatica*
(S: Diya-midella)

IDENTIFICATION Evergreen tree that grows to
10–20m tall. Pinkish-white flowers appear in racemes
borne at tips of branchlets. Flowers have four white
petals. Stamens pinkish-white, numerous and showy.
Leaves simple, leathery, shiny, stalkless (sessile) and
obovate in shape. Leaf arrangement is opposite.
Fruit is pyramidal in shape, four angled, and green
to brown in colour. Sepals remain at tips of fruits.
DISTRIBUTION Native. Coastal areas. **HABITAT**
Coastal areas. **FLOWERING PERIOD** May–August.

PRIMULACEAE (PRIMROSES)
The primrose family comprises about 1,000 species in 20 genera, many of which are familiar as cultivated plants for gardens. Most are found in temperate and mountainous regions of the northern hemisphere. Only one species occurs naturally south of the Equator, in South America. Some species are found in aquatic habitats.

Lunudan ▪ *Ardisia willisii*
(S: Lunudan)

IDENTIFICATION Shrubby plant that grows to about 2m tall. Leaves clustered at ends of branches, and are elliptic, thick, shiny and smooth (glabrous). Inflorescence is terminal, drooping, a simple or compound panicle. Inflorescence axis and flower peduncles maroon to dark brown. Flowers 0.6mm across. Five sepals reddish-brown, and five petals bright pink. Fruit is a drupe, a fleshy fruit with a hard seed inside. Drupes are 8–10mm in diameter, with edible fleshy pulp inside. Seeds have a fibrous coat. **DISTRIBUTION** Endemic. Wet-zones forests in south-west. **HABITAT** Grows as an undergrowth plant in lowland forests. **FLOWERING PERIOD** December–May.

> ### ERICACEAE (HEATHERS)
> This widely distributed family comprises about 3,500 species in 125 genera, which grow mainly as shrubs, and occasionally as trees. They include the small heathers of European heathland, as well as the shrubby evergreen rhododendrons of the Himalayas. The rhododendron found in Horton Plains grows as shrubs in exposed areas, and as trees in sheltered parts near cloud forests. South Africa is the centre of diversity for heathers, and an area including western China, Tibet, Myanmar and Assam for rhododendrons. These ericaceous plants generally prefer acidic soils and have a cosmopolitan distribution, but are absent from deserts.

Ceylon Rhododendron ▪ *Rhododendron arboreum zeylanicaum*
(S: Ma-rat mal)

IDENTIFICATION Eye-catching plant of the patana grassland and cloud forests of the Central Highlands. Grows to 1–2m in open plains, and to 3–5m in forests where it is sheltered from the wind. Leaves simple and stiff. Upper surface deep green, lower surface brown due to a dense mat of brown hairs. During the flowering season oval-shaped flower buds form at end of each branchlet. From time to time, deep red flowers arise from the buds. Each flower bunch contains 10–20 flowers at a time. Flower trumpet shaped with many stamens in centre. DISTRIBUTION Endemic. Horton Plains National Park, Loolkandura, Peak Wilderness Sanctuary, Nuwara Eliya, Pedro Nature Reserve and a few locations in the Knuckles Range. HABITAT Wet patana grassland and montane forests. FLOWERING PERIOD April–June.

Indian Cranberry ▪ *Vaccinium leschenaultia*
(S: Boralu)

IDENTIFICATION Shrub-like, much-branched small tree growing to about 1m tall. Bark grey. Leaves simple, elliptic to ovate, and stiff. Veins prominent on underside of leaf. Racemes are terminal or axillary, and there are a few flowers per raceme. Flower has five fused sepals, forming a greenish-red cup shape. Five bright pink petals are fused together, forming a corolla tube. Single flower is nearly 1cm in length. Stamens do not grow out over floral tube. DISTRIBUTION Native. Forested and open places of upper montane locations. HABITAT Restricted to upper montane forests. FLOWERING PERIOD February, March, September.

RUBIACEAE (COFFEE OR MADDERS)

This large family of more than 13,000 species in about 615 genera is the fourth largest family of flowering plants after the Orchidaceae, Asteraceae and Fabaceae. It is widespread and even found in the subpolar regions of the Arctic and Antarctic. The species diversity is greatest in the tropics and in humid tropical forests. Plants in the family are often the predominant woody species in humid tropical forests, and they grow as annual and perennial herbs, shrubs, lianas, epiphytes and trees. Some species (such as those in the genus *Myrmecodia*) have chambers in the nodes for ants. The simple, entire leaves are alternate or whorled. The inflorescences are variable, and the flowers are usually bisexual but also unisexual.

Spiny Randia ▪ *Catunaregam spinosa*
(T: Karai)

IDENTIFICATION Shrub that grows to about 5m tall. Spines conspicuous and 2cm long. Leaves simple, deep green, spoon shaped and clustered on dwarf branchlets on nodes. Sometimes leaf margins are rolled down. Flowers solitary. Five petals white, turning yellow with time. Five anthers in throat of a floral tube. Fruit 1–2cm in diameter, crowned with persistent calyx lobes. **DISTRIBUTION** Native. Dry lowlands, dry scrub and along the coast. **HABITAT** Dry scrubland. **FLOWERING PERIOD** June, August– September, November.

Flame of the Woods ■ *Ixora coccinea* var. *intermedia*
(S: Ratmal)

IDENTIFICATION Small tree or bush. Leaves elliptic, oblong or obovate, 3–14cm long, leaf tips acuminate, base heart shaped. Leaf arrangement is alternate opposite pairs of leaves at right angles to each other (decussate). Inflorescence is a cymose panicle with many red florets. Flowers usually grow in groups of three. Corolla lobes ovate or elliptic. Four tiny anthers, each one between a pair of petals. Style divided into two (bifid). **DISTRIBUTION** Exotic. Throughout Sri Lanka. **HABITAT** Cultivated plant grown in gardens. **FLOWERING PERIOD** Throughout the year.

Siamese White Ixora ■ *Ixora finlaysoniana*
(S: Suduratmal)

IDENTIFICATION Small tree. Bark fissured. Leaves oval shaped, 12–25cm long, stiff, tip acuminate. Inflorescence is terminal with many white florets. Each floret has four petals and a floral tube about 2–3cm long. Side margins of petals curl down. Four anthers, each located in between two petals. Stigma white, bifid and projects from floral tube. **DISTRIBUTION** Exotic. Wet and intermediate zones. **HABITAT** Cultivated plant found in gardens. **FLOWERING PERIOD** Throughout the year.

Knox Flower ■ *Knoxia hirsuta*

IDENTIFICATION Herb that grows to about 2m tall. Base of stem woody. Leaves simple and about 1.5–5cm long, with very clear veins. Flowers 0.8cm across; either stamens or style is projected over the petals. Sepals brownish-green, with coarse, stiff hairs. Four white petals. Pink bases of petals create pink ring around mouth of floral tube, which is densely covered with downy hairs (pubescence). **DISTRIBUTION** Endemic. Upper montane forests. **HABITAT** Common in forests and open places of upper montane forests. **FLOWERING PERIOD** Throughout the year.

Great Morinda ■ *Morinda citrifolia*
(S: Ahu; T: Manchavanna)

IDENTIFICATION Small tree that grows to about 6m tall. Leaves simple, large, about 10–30cm long, broadly elliptic shaped and shiny. Flowers about 1.3cm across, white and clustered in globular head. Throat hairy. Compound fruit formed by combination of fruits of several flowers. **DISTRIBUTION** Native. Along coast and up to mid-elevation areas. **HABITAT** Common in coastal areas and mangroves. Also cultivated in inland areas. **FLOWERING PERIOD** March, September–October, December.

White Mussanda ■ *Mussaenda frondosa*
(S: Mussenda, Muswenna, Wel-but-sarana; T: Vel-illai)

IDENTIFICATION Branchlets of this shrub covered with coarse, stiff hairs (hirsute). Leaves simple, large, about 12cm long, with both surfaces hirsute. Leaf veins deeply hairy. Flower tube about 2.7cm long, and hairy outside and inside, especially near mouth. Corolla bright orange with central yellowish rectangular patch. Sepals linear with downy hairs (pubescent). An unusual feature of this plant is that one sepal is large and pale, looking just like a leaf – most casual observers will simply assume that these sepals are whitish leaves. **DISTRIBUTION** Native. Moist locations up to 1,200m. **HABITAT** Common in secondary forests, and open and disturbed areas. **FLOWERING PERIOD** April–August.

Yellow Cheesewood ■ *Nauclea orientalis*
(S: Rata bakme, Bakme)

IDENTIFICATION Large tree that grows to about 30m tall. Bark dark brown, thick, fissured and flaky. Leaves simple, broadly ovate, about 30cm long, stiff, with blunt tips. Leaf

petiole distinct; about 4cm. Inflorescence is a solitary round head, and diameter of head with flowers is 3cm. Corolla yellow with four lobes. Styles conspicuous, 1–1.5cm long, white and project from corolla. **DISTRIBUTION** Native. Dry deciduous forests and secondary forests. **HABITAT** Moist places in dry-zone locations. Introduced to wet zone. **FLOWERING PERIOD** May.

Pleiocraterium plantaginifolium

IDENTIFICATION Perennial tussock-forming herb. Leaves simple, linear lanceolate, about 30cm long and form rosette. Leaf tip acute. Major veins parallel to midrib and to each other. Inflorescence is axillary, divided into many lateral branches, with the major axis up to 30cm long. Flowers light purple to white. Stamens are projected (exserted). **DISTRIBUTION** Endemic. Wet patana grassland in higher elevations. **HABITAT** Grows well in wet marshy grassland. **FLOWERING PERIOD** March.

LOGANIACEAE (LOGANIAS)
This diverse, widespread family comprises about 420 species in around 15 genera. They grow as annual and perennial herbs, woody climbers, lianas and large trees, and are found in a range of habitats, from arid areas to rainforests, mainly in the tropics and subtropics, extending to warm temperate areas. Their leaves are entire and opposite, with the petiole bases joined by a raised line. The inflorescence is usually cymose and terminal, and sometimes axillary. The flowers are radially symmetrical, and unisexual or bisexual. In some genera the plants bear either female or bisexual flowers (gynodioecious). The corolla is usually fused to form a narrow tube with short lobes. The Loganiaceae are closely related to the gentians. The family is divided into four tribes, one of which is the Strychnae, with plants that contain poisonous alkaloids from which strychnine is made. Some authors include the genus *Fagraea* in the Gentianaceae.

Perfume Flower Tree ■ *Fagraea ceilanica*
(S: Etamburu)

IDENTIFICATION Small, shrub-like tree that grows to about 3m tall. Sometimes grows as an epiphyte. Leaves fleshy, large (about 20cm long), and opposite. Leaf margin smooth, with midrib prominent on upper surface. Petiole very short. Flowers borne on terminal cymes, and are sweet scented, very large and pale yellow. Five green sepals. Corolla fleshy and funnel shaped. Stamens light brown and within floral tube. Style positioned higher than stamens. **DISTRIBUTION** Native. Primary and secondary forests from sea level to 1,500m or more. **HABITAT** Dry or marshy areas. **FLOWERING PERIOD** June.

GENTIANACEAE (GENTIANS)
Popular in cultivation, the gentians comprise more than 1,200 species in 75 genera, and occur from the subtropics, to temperate regions and mountains. A number of species have medicinal properties and are also used to provide a bitter flavour to food preparations. The herbaceous species often have four or five petals, and radially symmetrical flowers.

Binara ■ *Exacum trinervium* ssp. *macranthum*
(S: Binara)

IDENTIFICATION Herbaceous plant that grows to a height of 1m. Leaves simple, 3–8cm long and 3–5 veined. Five calyx lobes (fused sepals) green and winged, with wings rounded at base. Corolla lobes (petals) are spreading. In subspecies *E. t. pallidum* and *E. t. trinervium* (p. 138) the petals do not spread as they characteristically do in this subspecies. Petals violet to purplish-blue with pointed tips. Throat of floral tube yellow. Five yellow anthers. White style projects sideways away from stamens. **DISTRIBUTION** Endemic. Upper Central Highlands at 1,700–2,500m, for example in Horton Plains, Loolkandura and the Knuckles Range. **HABITAT** Occurs in wet patana vegetation in the highlands. **FLOWERING PERIOD** January–June, September–December.

Binara ▪ *Exacum trinervium ssp. pallidum*
(S: Binara)

IDENTIFICATION Subspecies with a slender stem, growing to about 1m tall. Stem quadrangular. Flowers showy (6cm across) and produced at apices of branches. Calyx formed of five green, winged sepals. Corolla lobes (or petals) ovate-elliptic to ovate-oblong, pale blue or dark blue. Throat of floral tube yellow. Five yellow anthers, abruptly narrowing at tips. **DISTRIBUTION** Endemic. Forests and grasslands of Central Highlands such as

the Knuckles Range. Occasionally present in Horton Plains National Park. **HABITAT** Sloping ground, often close to streams, in wet uplands. **FLOWERING PERIOD** February–March, September–December.

Binara ▪ *Exacum trinervium* ssp. *trinervium*
(S: Binara, Ginihiriya)

IDENTIFICATION Herb that grows to about 1m tall. Stem green and quadrangular. Simple leaves lack a petiole (sessile), and are stiff and three veined. Flower bunches grow at top of plant. Flower (6cm across) has five obovate petals. Five sepals green and winged. Petals generally blue to purple, although colour may vary according to location. Petals

form floral tube near flower stalks (pedicel). Throat of floral tube pale yellowish-green. Five stamens showy, yellow and gradually narrow towards tips. Style grows sideways and away from anthers. **DISTRIBUTION** Endemic. Low-country and intermediate zone to 200m. **HABITAT** Exposed roadside embankments, close to tributaries, and sloping grounds in wet lowlands. **FLOWERING PERIOD** Throughout the year.

White Binara ■ *Exacum walkeri*
(S: Sudu binara)

IDENTIFICATION Herbaceous plant about 50cm tall. Base of stem quadrangular. Leaves deep green, simple and ovate in shape. Leaf base heart shaped (cordate), and tip is pointed. Each leaf has three clear veins, and leaves are arranged oppositely. Solitary, showy flowers white, 5–7cm in diameter. Base of each petal yellow, and each flower has five petals and five yellow stamens. **DISTRIBUTION** Endemic. Highlands; sites include Horton Plains National Park, Nuwara Eliya and Pedro Nature Reserve. **HABITAT** Cold, wet areas of wet patana grassland and montane forests. **FLOWERING PERIOD** Throughout the year.

Gentiana quadrifaria

IDENTIFICATION Annual small herb that is hidden within tussock grasses and therefore difficult to find. Stem green, branched and quadrangular. Leaves numerous, very small (5mm long), and arranged in alternating opposite pairs at right angles to each other (opposite-decussate). Green leaves hard, with a purplish-brown margin, and leaves near tips of branches heavily crowded. Clear midrib appears only on vein of a leaf. Branches always terminate with indigo-blue to blue flower. Petals fused and 1cm across. Flowers open fully in good sunlight. **DISTRIBUTION** Native. Wet patana grassland in hill-country sites such as Horton Plains. **HABITAT** Wet patana grassland in montane zone. Prefers wet places with cold temperatures. **FLOWERING PERIOD** Throughout the year.

APOCYNACEAE (DOGBANES)

The dogbane family is predominant in the tropics and subtropics, although its range extends to temperate regions. In warm regions dogbanes occur mainly as woody plants (trees, shrubs and lianas), but in temperate regions they are found as perennial herbs. Some of the shrubs are spiny. There are more than 2,000 species in 200 genera, including the familiar Oleander *Nerium oleander*, which is extensively planted for its flowers. Many species are poisonous, being rich in alkaloids or glycosides.

Crown Flower ▪ *Calotropis gigantea*
(S: Wara)

IDENTIFICATION Large hardy shrub that grows to about 3–5m tall. Milky stems. Leaves leathery, large and oblanceolate in shape. Leaf arrangement is opposite. Leaves nearly lacking a petiole. Clusters of waxy, white or lavender flowers. Each consists of five pointed petals and small, elegant 'crown' rising from centre, which holds the stamens. Fruit dry and formed from single carpel, which splits along one side only (a follicle) and is recurved near stalk. Follicle about 6–10cm long. Inside of follicle woolly, and woolly hairs are white. **DISTRIBUTION** Native. Dry and arid zones and coastal belt of Sri Lanka. **HABITAT** Common in disturbed vegetation, and in waste places and abandoned areas, dry lake beds, bunds and man-made habitats. In the Middle East it can grow on sand dunes in deserts. **FLOWERING PERIOD** Throughout the year.

Carrissa ■ *Carrisa spinarum*
(S: Heen Koramba)

IDENTIFICATION Extensively spreading shrub with abundant thorns on stems (the 'spinarum' in the scientific name means 'of thorns'). Thorns sharp and nearly 2–3cm long. Leaves small (2–4cm long), oval, shiny, leathery and densely crowded on stem, with an opposite leaf arrangement. Flowers small (4–5cm in diameter), white, scented and star shaped. Each has five petals and floral tube at back of flower. Fruits are edible, ovate-shaped green berries, which turn blackish-purple when ripening. **DISTRIBUTION** Native. Mainly low-country dry-zone forests and jungles, roadsides and forest edges. Also dry locations of mid-country, montane forests and dry patana grassland. **HABITAT** Grows abundantly in semi-arid habitats with harsh environmental conditions with direct sunlight, and also on stony ground and in disturbed areas. **FLOWERING PERIOD** Flowers throughout the year, but mainly seasonal – flowers especially in February–May.

Madagascar Periwinkle ■ *Catharanthus roseus*
(S: Minimal)

IDENTIFICATION Ornamental plant rarely seen in the wild. Shrubby and herbaceous, and may grow to 1m tall. Leaves oval, nearly 7–9cm in length, with pale green midrib, and arranged on stem as opposite pairs. Plant produces many flowers at a time, and colour of flowers varies according to variety. Some varieties produce dark pink flowers with white centres, while one produces pale lavender flowers with dark violet centres. Pure white flowers are also possible in some varieties. Flower has five petals and a basal floral tube. Fruit is a narrow, cylinder-like follicle with many grooved seeds, and is green when young and turns brown with maturity.

DISTRIBUTION Exotic. Throughout Sri Lanka. Common in home gardens, botanical gardens and crematoria. **HABITAT** Dry places of upper hills, mid-country and dry-zone habitats. **FLOWERING PERIOD** Throughout the year.

Poison Nut ▪ *Pagiantha dichotoma*
(S: Kaduru)

IDENTIFICATION Grows to 6–7m tall. Young stems shiny. Leaves large, thick, hard and shiny, and leaf arrangement on stem is opposite. Secondary veins of leaves prominent and parallel to each other. Flowers borne on dichasially branched cymes. A dichasial cyme is where an inflorescence has a central stalk with a terminal flower. Laterally and opposite each other are two branches with a terminal flower, and again each of these has a pair of lateral and opposite branches bearing flowers. Flowers showy, each with five spirally arranged white petals. Bases of petals join together to form floral tube. Bases of petals and inside of floral tube yellow. Pendulous fruits large (10cm long), and green when young, turning to yellow to bright orange when maturing. The bright fruits are not eaten by birds, and parts of the plant are considered poisonous. Leaves and bark are used in

indigenous medicine, and wood for carving because it does not attract insects. **DISTRIBUTION** Native. Riverine habitats and lowland rainforests, mid-country wet forests and montane forests. **HABITAT** Common in shady moist places like riverine habitats. **FLOWERING PERIOD** Throughout the year.

Dense-Flowered Snake Root ▪ *Rauvolfia densiflora*

IDENTIFICATION Shrub-like plant that grows to about 2–3m tall. Stems narrow and white: they exude milky secretions when damaged. Leaves thin and spirally arranged over stem. Cluster of leaves at end of each branch. Flowers borne

in a terminal inflorescence. Each flower cluster bears more than 30 flowers. Flowers tiny (1cm in diameter), and in some plants white, but in others pinkish. Each flower has five petals. Bases of petals fused together, forming floral tube. Fruits ovoid in shape, small (1.5cm in diameter), and turn purple when ripening. **DISTRIBUTION** Native. Lower to mid-elevations in

lowland rainforests. Lower elevations of montane forests. **HABITAT** Moist places with moderate sunlight. **FLOWERING PERIOD** Early months of the year.

Yellow Oleander ■ *Thevetia peruviana*
(S: Kaneru)

IDENTIFICATION Small tree that grows to about 3–5m tall. Leaves hard, narrow and long (nearly 15cm in length), and spirally arranged and profuse on stem. They discharge a white latex. Flowers bright yellow and funnel shaped. Each has five spirally arranged petals. Bases of petals joined together to form floral tube. Fruits nearly 3–4cm in diameter, angled and nearly rhomboid in shape. Persistent calyx clearly visible on a fruit. Immature fruits green, turning brown to black on ripening. Usually two seeds inside each fruit. All parts of the plant, and especially the seeds, are highly poisonous. **DISTRIBUTION** Exotic. Home gardens and roadsides of lowland dry-zone locations. **HABITAT** Common in open, dry places with good sunlight. Often seen as hedge plants in dry zone. **FLOWERING PERIOD** Throughout the year.

> **CONVOLVULACEAE (BINDWEEDS)**
> The bindweeds are found in both the tropics and temperate regions. The family
> contains about 1,700 species in around 60 genera, most of which occur as clambering
> or twining herbs, although some are shrubs. Many species have cordate (heart-shaped)
> or arrow-shaped leaves, with single or paired, funnel-shaped flowers. Some bindweeds
> flower in the morning, hence the name 'morning glory'.

Argyreia osyrensis
(S: Dumbada)

IDENTIFICATION This plant can be considered as a liana. Mature stems woody, and young stems slender. Stems have whitish covering (indumentum). Leaves large (10cm long) and heart shaped. Underneath of leaves densely covered with a whitish indumentum. Inflorescences produced in terminal branches. Inflorescence axis, or peduncle, much longer than petioles, so flowers appear over the leaves. Numerous sepals in flowers covered with a dense indumentum. Flowers small (2cm in diameter), and pink to violet. Five petals of each flower fused together, forming funnel-shaped flower. Anthers projected over petals. Fruits small, shiny red and covered fully or partially by sepals. **DISTRIBUTION** Native. Common in disturbed areas of northeastern side of hill country. Also occurs in dry-zone locations such as Sigiriya. **HABITAT** Disturbed open land and roadsides. **FLOWERING PERIOD** Early months of the year.

Dwarf Morning Glory ■ *Evolvulus alsinoides*
(S: Vishnukranti)

IDENTIFICATION Perennial herbaceous creeper. Leaves small (2cm long) and elliptic shaped. They very nearly do not have a petiole and look stalkless (sessile). Alternate leaf arrangement is very clear. Leaves and stems covered with dense, white, shiny hairs. Each node produces a solitary flower. Flowers small (about 2cm in diameter) and blue. Each flower has five fused petals, and tip of each petal is bilobed. Centre of flower white, and five white anthers in each flower. Plant is widely used in traditional medicine. **DISTRIBUTION** Endemic. Seashores, lake boundaries and abandoned land in dry lowlands. **HABITAT** Sandy ground with good sunlight. **FLOWERING PERIOD** Throughout the year.

Swamp Morning Glory ■ *Ipomoea aquatica*
(S: Kankun)

IDENTIFICATION Vine that grows hugging the ground, attaching to the soil from each node. A hollow stem is one of the special features of this plant. Leaves have two triangular-shaped lobes at base (hastate). Leaf tip pointed, giving leaf an arrowhead shape. During the flowering period, a few purple flowers are produced in a cyme. Floral centre dark purple. Five petals of each flower fused together, forming funnel shape. Floral tube long, and contains anthers and ovary inside it. The plant is consumed all over the world as a leafy vegetable. **DISTRIBUTION** Exotic and cultivated. Wetlands all over Sri Lanka, but not very common in cold climatic areas of the Central Highlands. **HABITAT** Marshes, paddy fields, canals and wet, disturbed areas. **FLOWERING PERIOD** Throughout the year.

Railway Creeper ■ *Ipomoea cairica*

IDENTIFICATION Vine with twining stems that help the plant to climb over other plants living in the same habitat. Perennial with an underground rootstock. Leaves palmately

lobed, mainly into five lobes. When flowering, the plant produces inflorescences with many flowers, but there is only one flower in bloom at a time per inflorescence. Flowers (each nearly 6cm in diameter) purple, each with five fused petals. Long floral tube mounted on short flower stalk (pedicel). **DISTRIBUTION** Exotic. Throughout Sri Lanka except arid areas. **HABITAT** Common on cultivated land and wasteland, in thickets and hedges, and on stream banks, railway tracks and roadsides. **FLOWERING PERIOD** Flowers common in February–September. Flowering sometimes seen during other months.

Beach Morning Glory ■ *Ipomoea pes-caprae*
(S: Mudu bimthambaru)

IDENTIFICATION Perennial vine that grows on the grounds of sandy seashores. Stems long and trailing, with roots produced at each node. Leaves alternate, simple, thick, stiff and fleshy, with deeply bilobed leaf tips. Produces flowers on an inflorescence, with only one flower in bloom per inflorescence at a time. Flowers (about 6–8cm in diameter) pink to lavender-purple and funnel shaped, with dark purple throats. When the flowers are in bloom, the mat of green and purple on the ground is a picturesque sight. Five fused petals in each flower. Flowers open mainly in the morning, hence the name Beach Morning Glory. **DISTRIBUTION** Native. Commonly on coastal beaches and dunes all over Sri Lanka. **HABITAT** Sandy seashore habitats. **FLOWERING PERIOD** Throughout the year.

OLEACEAE (OLIVES)
This is a widespread family in tropical and temperate regions, with about 600 species in 25 genera, which grow as woody climbers, shrubs or trees. Some genera, such as *Fraxinus*, are widespread from the temperate north to the tropics of Costa Rica and Malaysia, but most tropical genera have a narrow distribution. Plants in this family that have economic use include the Olive *Olea europaea*, jasmines (*Jasminum*) and ash (*Fraxinus*) for timber, and garden plants such as lilacs (*Syringa*).

Wild Jasmine ▪ *Jasminum angustifolium*
(S: Wal-pichcha)

IDENTIFICATION Grows as a climber, and is sometimes a woody climber. Branches slender and long. Leaves simple, opposite and moderately stiff and leathery (subcoriaceous). Inflorescences are terminal and many flowered. Flowers 2–4cm across, pure white and with 7–9 petals. A floral tube is present. Green calyx tube has needle-like, pointed sepals. Fruit is a berry that turns black when ripe. **DISTRIBUTION** Native. Lowlands to 1,500m. **HABITAT** Common in dry habitats near streams. **FLOWERING PERIOD** June.

GESNERIACEAE (GLOXINAS)

The gloxinas comprise mainly tropical herbs and shrubs, with a few trees, and they form a fairly large family with about 3,000 species in 130 plus genera. They usually have opposite leaves. The inflorescence is generally a terminal compound cyme. The radially symmetrical flowers are usually dioecious, and occasionally unisexual, the petals are fused at the base in most species, and there are typically two (sometimes four) stamens.

Didymocarpus humboldtianus

IDENTIFICATION Small plant with an underground (rhizomatous) stem. Leaves large (20cm long), and all are attached to base of plant and arranged like a rosette. Leaf margin clearly notched with rounded teeth (crenate). Reticulate venation is noticeable on leaf, and young leaves have long, soft hairs (villous). Petiole is winged. At the time of flowering, 3–4 white flowers bloom in an inflorescence. Peduncle and sepals velvet in texture and villous. Flower white with yellow spot inside floral tube, and style extends from floral tube. **DISTRIBUTION** Native. Montane regions to about 1,700m. **HABITAT** Prefers moist, cool, shady places. Can be seen in rock crevices and on small pockets of soil on rock surfaces. **FLOWERING PERIOD** Generally throughout the year except in extreme dry periods. Most commonly seen in flower after monsoon rains.

LINDERNIACEAE (PIMPERNELS)
This family was previously treated as a part of the Scrophulariaceae. It comprises about 200 species in 16 genera, and is mainly neotropical. The plants are chiefly herbaceous annuals, but can be shrub-like. Their leaves are opposite, typically untoothed or with very fine teeth. The flowers are vertically symmetrical, and borne on terminal or lateral racemes. They are two lipped, bisexual and have five sepals. The ovary is superior, with two carpels and many seeds.

Rock Violet ▪ *Torenia cyanea*

IDENTIFICATION Plant with a quadrangular-shaped, succulent creeping stem. Leaves triangular in shape with purple-tinged midrib. Solitary flowers produced on long stalk. Corolla velvety, with colour varying from bluish to purple – the purple contrasts with patches of yellow in the base, making the plant very attractive. Four corolla lobes in each flower. **DISTRIBUTION** Endemic. Wet lowlands above 2,000m, and wet hill country to about 1,800m. **HABITAT** Prefers shady, wet and boggy places close to streams and similar places. **FLOWERING PERIOD** Throughout the year.

LAMIACEAE (DEADNETTLES OR MINTS)
The Lamiaceae, also known as the Labiatae, is a family of 230 genera with about 7,500 species, displaying a huge variety in how they grow. Many familiar herbaceous plants such as thymes (*Thymus*), Marjoram *Origanum majorana* and lavenders (*Lavandula*), are in this family, as well the large tropical hardwood, the Teak *Tectona grandis*, which is cultivated extensively for its timber. Other tropical trees in the genus *Vitex* are also included in the family. Many of the herbaceous species are aromatic and have hairy leaves, and the family has a worldwide distribution. The bisexual flowers are bilaterally symmetrical, with five united sepals and five united petals. The petals are typically fused to form an upper and lower lip, which gave rise to the old name Labiatae (from the Latin word *labia* for lip). The flowers are usually borne in two clusters that superficially look like a whorl of flowers. The leaves are opposite in pairs, with each pair at right angles to the previous pair.

Plectranthus kanneliyensis

IDENTIFICATION Perennial plant with succulent, quadrangular stems, in which the formation of the roots can be seen on the stem near the ground. Leaves large and pointed.

Leaf margin has shallow, rounded teeth (shallowly crenate). Midrib and petiole pinkish-red. Flowers borne in terminal panicle up to 10–11cm long, and appear in lateral branches. Flowers two lipped (bilabiate), drooping, and white and purple. Inflorescence axis pinkish-red. **DISTRIBUTION** Endemic. Wet lowland forests up to 200m. **HABITAT** Favours partial shade near streams and wet areas. The main walking trail in Sinharaja is a good place in which to see the plant. **FLOWERING PERIOD** June–September.

Lion's Ear ■ *Leonotis nepetifolia*
(S: Maha-yakwanassa; T: Kasitumpai, Perunthumbai)

IDENTIFICATION Large herb that grows to about 3m tall. Stem green, quadrangular, with downy hairs (pubescence). Opposite sides of stem deeply grooved. Leaves in upper areas of stem are small, while those in lower areas are large. They emit a characteristic smell when crushed. Inflorescences round (globose), and produced at each node spaced at intervals. Green, hairy bracts in inflorescences have spiny tips. A few blooms are present at a time on an inflorescence. Orange flowers covered with woolly hairs. **DISTRIBUTION** Exotic.

Dry and intermediate regions in low country to 500m. **HABITAT** Wasteland and roadsides in dry locations. Sometimes grows in wet zone. Common roadside plant in dry-zone national parks. **FLOWERING PERIOD** November–June.

Ceylon Slitwort ■ *Leucas zeylanica*
(S: Geta-tumba; T: Mudi-tumpai)

IDENTIFICATION Much-branched annual herb with green quadrangular stems. Leaves linear, deep green. Plant parts emit characteristic smell when crushed. Flowers borne on terminal heads, with 2–3 whorls (verticils) of flowers. Two to three white, two-lipped (bilabiate) flowers present at a time. Flower small, 0.4cm across x 1.5cm long. **DISTRIBUTION** Native. Dry and wet lowlands and uplands from sea level to 1,700m. **HABITAT** Roadsides, wasteland and sandy places. **FLOWERING PERIOD** Throughout the year.

Plectranthus nigrescens

IDENTIFICATION Perennial herb with stem about 30cm high, slender and distinctly four angular. Leaves ovate. Racemes terminal, about 20cm long; 6–10 flowers. Flowers small, 1–1.5cm long. Corolla tubular, two lipped (bilobed) and purple. Centre of large floral lobe has patch of white. Underside of corolla crowded with purple to lavender hairs. Sepals green mixed with deep purple. Sometimes the axis is purple. **DISTRIBUTION** Native. Montane and submontane locations at about 1,000–2,700m. **HABITAT** Grows along shady roadsides, beneath montane and submontane forests, often close to streams. **FLOWERING PERIOD** Throughout the year.

OROBANCHACEAE (BROOMRAPES)
This family was once treated as part of the Scrophulariaceae family. All but three of its genera are fully or partially parasitic on other plants. Species that are fully parasitic lack leaves with chlorophyll for photosynthesis and are root parasites. The family is widespread, with a worldwide distribution extending to the subarctic, and contains about 2,000 species in around 90 genera.

Christisonia lawii

IDENTIFICATION It is only when this plant is in flower that this species can be identified as a member of the genus *Christisonia*. This is because leaves are absent in this genus, which comprises plants that are root parasites. The flowers bloom on the forest floor because the leafless flower stalk (scape) is very short or absent. Calyx pale brownish-pink and has five lobes. Floral tube white. Corolla lobes dark blue-purple. Yellow patch on lower floral lobe. **DISTRIBUTION** Native. Middle elevations of mixed evergreen forests. **HABITAT** Parasite and root associate of plants in evergreen forests. **FLOWERING PERIOD** Photographed in October. Data not available.

Christisonia tricolor

IDENTIFICATION Like in the previous species, leaves are absent in this species. The flowers arise from a leafless stalk (scape) about 2.5–5cm tall, so they are only just above the forest floor. Calyx purplish-red and tubular. Corolla 5–7cm long. Outer surface of floral tube and undersides rosy-purple. Inside of floral tube yellow. Outside of floral lobes white with long, soft hairs (villous). **DISTRIBUTION** Endemic. Lower montane forests to about 1,500m. **HABITAT** Parasite on roots of plants in Acanthaceae family and bamboos. **FLOWERING PERIOD** July.

Bignoniaceae (Bignonias)

This family is found mainly in the tropics and subtropics, with the species diversity being highest in the neotropics. There are about 850 species in around 110 genera, and the family comprises climbers, lianas, shrubs and trees. The plants are adapted to a variety of habitats, from montane grassland to wet forests. Their leaves are compound and opposite with tendrils, and the leaf arrangement is usually opposite, but can be alternate or in whorls. The flowers are solitary or borne in axillary or terminal racemes. The calyx has five fused petals, and the stamens are usually attached to the corolla tube.

African Tulip Tree ▪ *Spathodea campanulata*
(S: Kudella)

IDENTIFICATION Large tree planted for its ornamental value. At the flowering time the crown is showy. Leaves compound, with 13–19 leaflets to a leaf. Inflorescences produced at ends of branchlets. Calyx golden-brown and velvety, and before opening is filled with liquid – when pressed, it squirts liquid for a considerable distance, and the flowers are used by children to play with. Flowers large (10cm in diameter), bright red to orange. Petals

of each flower joined together, forming a tube. Petal margin lined with yellow. Fruit is a capsule 15–23cm long, and brownish-black. **DISTRIBUTION** Exotic. Wet lowland and montane habitats from sea level to 1,200m or higher. Grown in home gardens, roadsides, and botanical and recreational gardens. Also abundant in tea-cultivation areas. **HABITAT** Prefers moist lowlands and montane regions. **FLOWERING PERIOD** June–September.

ACANTHACEAE (ACANTHUS)
Plants in this family of 3,500 species occur mainly as herbs, twining vines or shrubs. About 250 acanthus genera are found mainly in tropical and subtropical regions, with a few species occurring in temperate regions. The flowers are typically subtended by prominent bracts, which may be colourful and showy. Some species have medicinal properties. A number are spiny. The leaves are usually opposite, compound and sometimes have tendrils.

Hygrophila schulli
(S: Neeramulliya)

IDENTIFICATION Unbranched, hairy, shrubby herbaceous plant. Flowers, leaves and spines arranged as a whorl on a node. Leaves simple and deep green. Leaf margin wavy (undulate). Leaves and stem almost covered by prominent whitish hairs. Presence of spines is a characteristic feature of the plant, and they are brown and sharp. Flowers bright purple and have five petals that are arranged into two lobes. Upper lobe contains two petals, lower lobe contains three. Centre petal of lower lobe has characteristic yellow patch. Four yellowish-purple stamens – two stamens long, other two shorter. All stamens slightly curved towards centre of flower. **DISTRIBUTION** Native. Sigiriya, Habarana, Kandy and Matale. **HABITAT** Shallow, marshy land and roadside channels of dry-zone locations. Occasionally seen in wet zone in lowland and montane locations. **FLOWERING PERIOD** March, April.

Paper Plume ▪ *Justicia betonica*
(S: Sudupuruk)

IDENTIFICATION Perennial branched herb just over 1m tall. Stems erect and tinted with purple lines. Leaves green and arranged oppositely on stem. Sparse hairs at leaf margins and on veins. Inflorescence is a terminal spike. Spikes crowded with white, green-veined, leaf-like bracts, and flowers are tiny and hidden inside these bracts, with each flower being surrounded by three bracts. Flowers pinkish and lobed. Interior area of each flower purple. **DISTRIBUTION** Native. Easy to see all over Sri Lanka, including in lowlands, mid-hills and hill country. **HABITAT** Has begun to spread rapidly in disturbed, moderately wet, marshy lands, and is abundant near paddy fields, tributaries and riversides. **FLOWERING PERIOD** Throughout the year.

Susan Vine ■ *Thunbergia alata*
(S: Susange-esa)

IDENTIFICATION Herbaceous, annual or perennial, twining vine that spreads more than 3–4m away from the point of attachment to the soil. Leaves arrowhead shaped and hairy, and leaf margins irregularly toothed. Flowers attractive, with five petals, and typically warm orange with a characteristic chocolate-coloured dark spot in centre. Several colour varieties recorded throughout Sri Lanka – some light orange with chocolate-coloured central spot, others white with characteristic central spot of contrasting colour; occasionally flowers do not have chocolate-coloured central spot. Flowers borne singly in leaf axils. Two large green bracts at floral base. Floral tube and tiny, needle-like sepals in calyx can be seen if flower bracts are removed. **DISTRIBUTION** Exotic. Throughout Sri Lanka. Common in urban areas, for example in water channels that run through cities. **HABITAT** Needs moist conditions with direct sunlight. Best growth seen in disturbed areas, and channels where refuse and waste accumulate on waysides. Commonly seen in tea-cultivation areas at middle and high elevations. **FLOWERING PERIOD** Throughout the year.

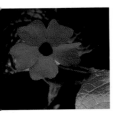  

White Lady ■ *Thunbergia fragrans*

IDENTIFICATION Perennial twining herb that climbs over low vegetation. Leaves dark green, nearly triangular in shape, hairy, and arranged oppositely on hairy stem. Flowers

solitary and borne on leaf axils. They are pure white and lack any fragrance despite the Latin epithet *fragrans*. Each flower has five petals and central, hollow, light green tube. Petals nearly square in shape. Two heart-shaped green bracts at base of each flower. Normally, flowers wilt later in the day. **DISTRIBUTION** Native. Widespread in wet zone from lowlands to highlands. Abundant on disturbed land and rarely seen in undisturbed places. **HABITAT** Grows well in wasteland and on roadsides. Also in home gardens. **FLOWERING PERIOD** Throughout the year.

> ### LENTIBULARIACEAE (BUTTERWORTS & BLADDERWORTS)
> This widespread family of carnivorous herbs comprises about 320 species in three genera. The plants are usually found in wet habitats, with some species growing partially submerged in water. The three genera, *Pinguicula*, *Genlisea* and *Utricularia*, are very different in how the carnivorous parts work. In *Pinguicula* the leaves secrete a mucilage that traps animals. The *Genlisea* have normal leaves in a rosette and the carnivorous leaves below them. The *Utricularia* (bladderworts) have complicated traps with doors that are usually borne on the stems or leaves. The flowers are borne on a leafless flower stalk (scape) arising from a rosette of leaves.

Blue Bladderwort ▪ *Utricularia caerulea*
(S: Nil monerrassa)

IDENTIFICATION Very small to medium-sized carnivorous plant that grows to less than 30cm tall and has a slender green stem. Inflorescence is on a terminal spike, with 3–4 flowers per spike. Flower small (about 0.8cm long). Sepal at back (posterior) brownish-pink. Sepals in front (anterior) greenish-brown and form tail behind flower. Corolla two lipped (bilobed) and purple. Large lower lip inverted and cup shaped. Colourful veins clear on lower lobe of corolla. **DISTRIBUTION** Native. Wet lowlands and montane areas to about 2,000m. **HABITAT** Grows as a terrestrial in wet, shallow soil over rocks, wet grassland and swamps, or near streams in open communities. **FLOWERING PERIOD** Photographed in March and April. Believed to be throughout the year, but detailed information on flowering not available.

Striped Bladderwort ▪ *Utricularia striatula*
(S: Nil monerrassa)

IDENTIFICATION Carnivorous plant that is a tiny herb. Flower 0.6cm across. Sepals reddish-pink. Corolla violet with large yellow spot at base of lower lip. Margin of lower lip divided into five lobes. **DISTRIBUTION** Native. Montane forests. **HABITAT** Wet rocks, wet, moss-covered tree trunks, under partial shade. **FLOWERING PERIOD** Photographed in July.

CAMPANULACEAE (BELLFLOWERS & LOBELIAS)
This very diverse group of plants varies from the typical 'bellflowers', to plants such as lobelias, and some species are aquatic. There are more than 2,000 species, mainly in northern temperate zones. They grow as herbs and shrubs, and a few are small trees with milky sap. The corolla of their flower is lobed, often forming a tube, and the flowers are bisexual. The bellflowers are typically blue, and are popular garden plants. The showy flowers are designed to attract insect pollinators. The fruits are typically in the form of berries.

Southern Rockbell ■ *Wahlenbergia marginata*

IDENTIFICATION Small (30cm tall) perennial herb. Leaves small (1–2cm long) and spirally arranged. Flowers (1cm in diameter) solitary. Flower stalk can be as long as 10cm.

Flowers tend to open towards the ground. Five ashy-blue petals in each flower, and bases of petals are white. Flower-like white stigma in centre of each flower. **DISTRIBUTION** Native. Wet patana grassland of highlands at 1,400–2,400m. **HABITAT** Often found growing with sedges. Normally occurs in cold climatic conditions. **FLOWERING PERIOD** Flowers almost throughout the year, but chiefly in December–May.

ASTERACEAE (DAISIES)
The Asteraceae comprise the largest family of flowering plants, with about 23,000 species in more than 1,500 genera. They are also referred to as the Compositae, and occur mainly as herbaceous plants (such as the familiar daisies), as well as shrubs and rarely as trees. The flower (or what appears to be the flower) is characteristically a composite flower or capitulum, which is formed from a tight cluster of florets. It superficially looks like a single flower, but a daisy is not a single flower but a cluster of flowers.

Billy Goat Weed ■ *Ageratum conyzoides*

IDENTIFICATION Herbaceous plant that grows to 0.5–1m tall. Leaves ovate and nearly 3–6cm in length. Leaf arrangement is opposite. Stems and leaves covered with fine white hairs. Flowers white to mauve, and less than 6mm across. Terminal inflorescence is in form of a head of flowers with about 30–50 flowers in each head. Strong smell emitted when leaves or flowers are crushed. **DISTRIBUTION** Exotic. Throughout Sri Lanka. **HABITAT** Very common weed in Sri Lanka that commonly grows close to human habitation. Thrives in gardens and on agricultural land, and often found on disturbed sites and in degraded areas. Invades forests, woodland, grassland and cultivated land. **FLOWERING PERIOD** Throughout the year.

Anaphalis subdecurrens
(S: Sudana)

IDENTIFICATION Succulent, erect herbaceous plant that grows to about 1m tall. Leaves narrow and long, lack stalk (sessile), and arranged in spiral on stem. Stems and leaves greenish to ash in colour due to covering of dense, white, smoke-like, woolly hairs. Inflorescences are heads composed of congested cymes. Corolla pink, yellow or whitish-pink. **DISTRIBUTION** Native. Common in wet patana grassland in the Central Highlands. Also seen along roadsides of wet montane forests, and along roadsides of tea plantations at high altitudes. **HABITAT** Places with high moisture and low temperatures. Commonly grows in water bogs and peat in wet patana grassland. **FLOWERING PERIOD** Latter months of the year.

Bidens biternata

IDENTIFICATION Erect annual herb that grows to 1m tall. Leaves compound with three leaflets (trifoliate). Flowers are terminal heads that consist of yellow disc florets in the middle surrounded by white ray florets. Ray florets look like petals and disc florets look like stamens. Disc florets are numerous, whereas ray florets occur in a small, countable number. Black fruits are single seeded, dry and have a thin wall, not splitting when ripe (an achene). If you walk through these plants, the achenes stick to your clothes and are difficult to remove. This illustrates how they stick to the fur of mammals

and the feathers of birds, to be transported away from the mother plant. **DISTRIBUTION** Native. Throughout Sri Lanka. **HABITAT** Roadsides and wasteland. Also abandoned areas. **FLOWERING PERIOD** Throughout the year, but flowering abundancy is high in September–November.

Red Flower Rag Leaf ▪ *Crassocephalum crepidiodes*

IDENTIFICATION Common weed that is a herb with a succulent, branched stem. Grows to about 1m tall. Leaves large (nearly 10–15cm long), elliptic to oblanceolate in shape. Leaf margin irregularly toothed. Tiny white hairs on both surfaces of leaves, which can be felt by lightly touching the leaves. Petiole long and succulent. Leaves fragile, wilting temporarily and conspicuously when the sun rises. Floral stalk curves down towards the ground at noon. During the very early stages, flower head is pendulous; it becomes erect when it is maturing. Numerous tiny flowers in the head enveloped by whorl of green bracts (an involucre), which is cylindrical in shape. Flowers (florets) red. Fruit is an achene with a brown base. Series of white bristles on base of the corolla, which is later found at tip of fruit (a pappus). Fruits disperse through wind. **DISTRIBUTION** Exotic. Throughout Sri Lanka. **HABITAT** Cultivated areas with vegetables, paddy, maize and other plants. Also common along roadsides, in disturbed patches in forests and in many abandoned areas. **FLOWERING PERIOD** Throughout the year, but peaks in October–August.

Cupid's Shaving Brush ▪ *Emilia sonchifolia*
(S: Kadupahara)

IDENTIFICATION Erect annual herb that grows to 1m tall. Usually produces prostrate branches from very base of plant. Stems weak, greenish-purple in colour, although stem colour may vary according to location. Stems covered with tiny white hairs. Leaves green and alternate. Lower leaves have narrowly winged petioles and fine, blunt teeth, or deep, rounded pinnate lobes, and are broadly triangular. Upper leaves clasp stems and are smaller, sessile and sometimes coarsely toothed. Flower heads small, narrowly urn shaped to bell shaped. Florets are disc florets; there are no ray florets. Bases of florets covered by cylindrical, purplish-green envelope. Bloomed florets showy, light pink to lavender in colour. Flowers followed by seed heads with white, fluffy, wind-borne seeds. Leaves and stems emanate characteristic smell when crushed. Plant is widely used in traditional medicine. **DISTRIBUTION** Native. Throughout Sri Lanka. **HABITAT** Considered a weed and occurs along roadsides, and in wasteland and disturbed forest sites. Also found in areas with agricultural cultivation, among vegetables and paddy. Normally shows healthy growth in wet habitats. **FLOWERING PERIOD** Throughout the year.

Emilia zeylanica
(S: Pupula)

IDENTIFICATION Erect, firm herb that grows to 1–1.5m tall. Stem branched near base. Leaves sessile. Leaf bases auriculate. Leaf base broad and leaf blade narrows towards tip. Leaves densely covered by tiny pubescence. Inflorescence is a head 1–2cm in diameter. Florets in heads enveloped by cylindrical whorl of bracts beneath an inflorescence (cylindrical involucre). Pale purple florets strongly projected (exserted) from involucre. Fruit is an achene. Series of bristles (pappus) on fruit, which is prone to fall off easily (cauducous). Fruit dispersed by wind. **DISTRIBUTION** Endemic. Commonly seen in wet

patana grassland, on forest borders and on ridges at higher elevations. Occurs in roadsides in highlands, especially in the central hills. **HABITAT** Favours cold climates, wet soil and open areas. **FLOWERING PERIOD** Throughout the year.

Santa Barbara Daisy ▪ *Erigeron karvinskianus*

IDENTIFICATION Stems of this plant grow on the soil surface (prostrate stems). Stem is rooted at each node, and from the nodes the plant branches densely, forming a mat of vegetation. Leaves small, narrow, elliptic and lobed. Leaf-tip apex sharply pointed or tapering to sharp tip (acuminate). Both surfaces of leaf covered by tiny hairs. Inflorescence is a head and is produced on upper leaf axils. Head hemispherical and small (6–8mm in diameter), and stalk long (about 4–10cm). Both ray and disc florets on head. Disc florets yellow, numerous and in middle of head. Ray florets fewer and surround disc florets, forming outer circle. Ray florets white at first, turning purplish when wilting with age. White and purple flowers may occur at the same time on the same plant. **DISTRIBUTION** Exotic. Forests and human habitation in upper elevations of Sri Lanka. Plant provides good groundcover and assists in controlling soil erosion. **HABITAT** Common in roadside banks and water channels, and roadsides of disturbed forests in mountain areas. Also near tea plantations. Prefers moderately cold climates. **FLOWERING PERIOD** Throughout the year.

Small Flowered Galinsoga ■ *Galinsoga parviflora*

IDENTIFICATION Weed that grows to about 15–75cm tall. Leaves ovate and petiole nearly 1.5cm long. Leaf tip pointed. Both surfaces of leaves sparsely covered by hairs. Inflorescence is a head 0.7–1cm in diameter. Peduncle long (about 3cm). Middle yellow area of head is like a button. Nearly 30 disc florets present in centre. Each has five petals and five sepals. Sepals and petals of disc florets yellow. Four to five white ray florets per head. Three segments are joined together in ray florets. Fruit is a black achene. **DISTRIBUTION** Exotic. Throughout Sri Lanka. Common on cultivated land and wasteland. **HABITAT** Wet places with moderate sunlight. **FLOWERING PERIOD** Throughout the year.

Senecio ludens

IDENTIFICATION Perennial herb that grows to 15–30cm. Weak stem. Lower leaves have petioles loosely circular in shape. Upper leaves lack petioles (sessile), and bases have a flexible joint (that is, they are articulated). Leaf margins of both leaf types have rounded teeth (dentate). All leaf veins arise from leaf base and are directed towards margin. Both surfaces of leaf have long, soft hairs (villous), and leaves are spirally arranged around stem. Inflorescence is a yellow, showy head (1.5–2cm in diameter). A few heads produced on axis of each plant. Surrounding the head are 6–8 yellow ray florets, tips of which are bilobed or trilobed. About 40 small disc florets in centre. Fruit is an achene, covered by grey woolly hairs (pubescence). Pappus brown. **DISTRIBUTION** Native. Mainly undisturbed forest types in higher elevations in Sri Lanka, but seen along roadsides in some areas such as the Knuckles. **HABITAT** Common in moist conditions, and also found along roadsides, although its natural habitat is patana and forest scrub. **FLOWERING PERIOD** Throughout the year. Flowering intensity highest in October–April.

Sow Thistle ■ *Sonchus oleraceus*
(S: Wal dunkola)

IDENTIFICATION Annual herb that grows to nearly 1m tall. Shapes of leaves varied. Basal leaves ovate and small. Mid-leaves long (nearly 20cm) and have 1–3 pairs of lateral segments. Leaf margins have sharp teeth. Leaves lack a petiole. Sometimes there is an ear-shaped projection (auricle) at the leaf base. Leaves arranged spirally. Like other composites, this plant produces heads (1.5cm in diameter) as inflorescences. All florets are ray type (disc florets are absent). However, inner florets are smaller than outer ones. Florets yellow. All florets enveloped in green cylindrical whorl of bracts (involucre), and each floret

produces a brown achene. Pappus white and 7–8mm long. **DISTRIBUTION** Exotic. Submontane and montane areas of Sri Lanka. **HABITAT** Roadsides and agricultural land with moderate sunlight. **FLOWERING PERIOD** Throughout the year.

Dandelion ■ *Taraxacum javanicum*

IDENTIFICATION Ground-hugging plant in which stem cannot be seen. Leaves arise and spread, radiating on the ground. Leaves simple, oblanceolate, 5–15cm long, tapering at base into petiole. Leaf margin irregularly toothed. Green leaf has 3–4 pairs of lateral lobes. Terminal lobe triangular (deltate) in shape. Middle frond, together with petiole, reddish-brown. Only one inflorescence (4–5cm in diameter) at a time, which grows above the ground. Leafless part of stalk of inflorescence (scape) 15–25cm tall, and reddish-brown. Inflorescence consists of ray florets. Outer florets spread laterally, and inner ray florets erect and small in comparison to outer ones. Outermost (distal) ends of florets lobed into five parts. All florets yellow. Achenes and white seed parachutes form characteristic white dandelion seed head when seeds are ready for dispersal by wind. **DISTRIBUTION** Exotic.

Highlands at about 2,000m. Common in grassland and home gardens. **HABITAT** Prefers a cool climate. Commonly present with grasses. Also found on lawns in central hills. **FLOWERING PERIOD** January–August.

Wild Sunflower ■ *Tithonia diversifolia*
(S: Wal suriyakantha)

IDENTIFICATION Perennial herb that grows to 2–3m tall. Weak stem is branched. Leaf arrangement alternate, and leaves in lower part of branches are 3–5 lobed; leaves in upper part of stem not lobed. Leaves oblanceolate or triangular in shape. Winged petioles slender and as long as 2–10cm. Inflorescence head (capitula) yellow and nearly 10cm in diameter. Peduncle 10–25cm long. There are 10–15 ray florets per head, the ends of which are lobed, with 2–3 segments, and are curved towards the ground. Numerous disc florets in centre dark yellow or yellowish-orange. Seeds (achenes) brown. Leaves of plant are used as fodder and fertilizer in organic agriculture. **DISTRIBUTION** Exotic. Mid-elevations in wet, dry and submontane locations. **HABITAT** Roadsides, wasteland, disturbed areas and open ground with moderate sunlight. **FLOWERING PERIOD** April–July.

Tridax Daisy ■ *Tridax procumbens*

IDENTIFICATION Perennial herb with prostrate stem. Leaves simple and generally arrow shaped. Two leafy segments at base of leaf. Flower head (capitulum) 1.2cm in diameter and solitary on erect peduncle (10–20cm long). Whorl of bracts on flower (involucre) very clear. Ray florets ovate in shape, white and with 5–7 florets in each flower head. Ends of ray florets trilobed. Disc florets in centre yellow. Achenes blackish. **DISTRIBUTION** Exotic. Throughout Sri Lanka. **HABITAT** Prefers intense light and moderately high temperatures. Common along roadsides, and on lawns and cultivated ground. **FLOWERING PERIOD** Throughout the year.

Iron Weed ■ *Vernonia cinerea*
(S: Monara kudumbiya)

IDENTIFICATION Erect annual herb that grows to 15–30cm tall. Stem covered with tiny hairs, green, and violet to purple near nodes. Leaves ovate. Petiole long and winged. Leaf

arrangement around stem is spiral. Several flower heads (capitula) produced on a corymb. All florets in flower head are alike, purple or mauve, and each has 5–6 petals. Flower head covered by greenish-purple cylindrical whorls of bracts (involucre). Achenes dark brown or blackish. Pappus white. A characteristic strong smell is emitted when the plant parts are crushed. Plant is widely used in traditional medicine. **DISTRIBUTION** Native. Throughout Sri Lanka. **HABITAT** Common weed along roadsides and in wasteland. Can be found in harsh environmental conditions. **FLOWERING PERIOD** Throughout the year.

Golden Daisy ■ *Vicoa indica*

IDENTIFICATION Annual herb. Stem erect, growing up to 1m tall, cylindrical and reddish-brown. Leaves simple, triangular and stalkless. Base of leaf has ear-like projections. Both surfaces of leaves sparsely covered by soft hairs, and leaves are arranged spirally on stem. Flower head nearly 1.5cm in diameter. Peduncle slender. Numerous linear yellow ray florets around head. Tips of ray florets trilobed. Disc florets tubular and yellow, and small in size. Achene brown, pappus white. **DISTRIBUTION** Native. Dry-zone lowlands. Very common in northern dry areas of Sri Lanka; also in abandoned paddy fields. **HABITAT** Open land such as roadsides and cleared forest patches. **FLOWERING PERIOD** Throughout the year.

BIBLIOGRAPHY

Abeywickrama, B. A. (1959). *A Provisional Checklist of the Flowering Plants of Ceylon.* Ceylon Journal of Science 2(2): 119–240.

Abeywickrama, B. A. (1982). *The Families of the Flowering Plants of Sri Lanka – Part 1.* National Science Council of Sri Lanka, Colombo. p. 86.

Ashton, M. S., Gunatilleke, S., de Zoysa, N., et al. (1997). *A Field Guide to the Common Trees and Shrubs of Sri Lanka.* WHT Publications (Pvt) Ltd, Colombo. p. 432.

Beentje, H. (2016). *The Kew Plant Glossary: An Illustrated Dictionary of Plant Terms.* Kew Publishing, Royal Botanic Gardens, Kew. p. 184.

Bond, Thomas E. T. (1953). *Wild Flowers of the Ceylon Hills.* Geoffrey Cumberlage, Oxford University Press, Madras. p. 240.

Cramer, L. H. (1993). *A Forest Arboretum in the Dry Zone.* Institute of Fundamental Studies, Kandy. p. 243.

Dassanayake, M. D. & Fosberg, F. R. (eds). *A Revised Handbook to the Flora of Ceylon.* Vols 1–7, published between 1980 and 1991.

Dassanayake, M. D., Fosberg, F. R. & Clayton, W.D. (eds). *A Revised Handbook to the Flora of Ceylon.* Vols 8–10, published between 1991 and 1996.

Dassanayake, M. D. & Clayton, W. D. (eds). *A Revised Handbook to the Flora of Ceylon.* Vols 11–14, published between 1996 and 2000.

Dassanayake, M. D. & Shaffer-Fehre, M. (eds). *A Revised Handbook to the Flora of Ceylon.* Vol. 15, Parts A and B: Ferns and Ferns-Allies. Published in 2006.

de Vlas, J. & de Vlas-de Jong, J. (2008). *Illustrated Field Guide to the Flowers of Sri Lanka.* Mark Booksellers and Distributors (Pvt) Ltd: Sri Lanka. p. 269.

de Vlas, J. & de Vlas-de Jong, J. (2014). *Illustrated Field Guide to the Flowers of Sri Lanka. Volume 2.* Published by the authors: Netherlands. p. 320.

Fernando, D. (1980). *Wild Flowers of Ceylon.* Lake House Printers and Publishers Ltd., Colombo. p. 47.

Fernando, M., Wijesundara, S. & Fernando, S. (2003). *Orchids of Sri Lanka: A Conservationist's Companion.* IUCN, Sri Lanka. p. 147.

Gunatilleke, C. V. S. (2007). *A Nature Guide to the World's End Trail, Horton Plains.* 1st edn (reprint), Science Education Unit, Faculty of Science, University of Peradeniya. p. 63.

Heywood, V. H., Brummitt, R. K., Chulam, A. & Seberg, O. (2007). *Flowering Plant Families of the World.* Firefly Books. p. 424.

Heywood, V., Brummitt, R. K., Chulam, A. & Seberg, O. (2015). *Flowering Plants: A Pictorial Guide to the World's Flora.* Chartwell Books: New York, USA. p. 288.

Isaac-Williams, M. L. (1988). *An Introduction to the Orchids of Asia.* Angus & Robertson Publishers: UK. p. 262.

Jayaweera, D. M. A (1980). *Medicinal Plants (Indigenous and Exotic) Used in Ceylon.* The National Science Council of Sri Lanka, Colombo. 279 pages.

Jones, M. (1991). *Wild Flowers of Britain and Northern Europe.* Edn by Rainbow Books: London. First published 1980. p. 125.

Kehimkar, I. (2000). *Common Indian Wildflowers.* Bombay Natural History Society and Oxford University Press: India. p. 141.

Kottegoda, S. R. (1994). *Flowers of Sri Lanka*. The Royal Asiatic Society of Sri Lanka. p. 247.

Miththapala, S. & Miththapala, P. A. (1998). *What Tree is That? A Lay-person's Guide to Some Trees of Sri Lanka*. Ruk Rakaganno, Colombo. 93 pages.

Pemadasa, A. (1996). *The Green Mantle of Sri Lanka*. Sri Lanka National Library Services Board. p. 242.

Pethiyagoda, R. & Rodrigo, R. (1993). *A Provisional Index to a Revised Handbook to the Flora of Ceylon*. The Wildlife Heritage Trust of Sri Lanka. 251 pages.

Pethiyagoda, R. (2007). *Pearls, Spices and Green Gold*. WHT Publications (Private) Limited, Colombo. 241 pages.

Phillips, A. & Lamb, A. (1996). *Pitcher-Plants of Borneo*. Natural History Publication (Borneo) Sdn. Bhd. Kota Kinabalu. p. 171.

Ratnayake, H. D. & Ekanayake, S. P. (1995). *Common Wayside Trees of Sri Lanka*. Royal Botanical Gardens, Peradeniya. p. 192.

Senaratna, L. K. (2001). *A Checklist of the Flowering Plants of Sri Lanka*. MAB Check List and Hand Book Series Publication No. 22. National Science Foundation of Sri Lanka. p. 451.

Stevens, P. F. (2001 onwards). *Angiosperm Phylogeny Website*. Version 14, July 2017. www.mobot.org/MOBOT/research/Apweb. Accessed November 2017.

Sumithrarachi, D. B., Ratnayake, H. D. & Ekanayake, S. P. (1995). *Beautiful Wild Flowers of Sri Lanka*. Royal Botanical Gardens, Peradeniya. p. 154.

Trimen, H., Hooker, J. D. & Alston, A. H. G. (1893 to 1931). *Handbook of the Flora of Ceylon*. Parts 1 to 5 plus supplement. Dulau & Co.: London.

Werner, W. (2001). *Sri Lanka's Magnificent Cloud Forests*. WHT Publications, Colombo. p. 96.

Worthington, T. B. (1950). *Ceylon Trees*. The Colombo Apothecaries' Co. Ltd. p. 429.

RESOURCES

ORGANIZATIONS

This list of organizations includes what may appear to be bird-centric or animal-centric bodies. However, their field meetings provide a good opportunity to get out into decent plant sites and meet people who know plants.

The Sri Lanka Natural History Society (SLNHS)

www.slnhs.lk. Email: muhudubella@gmail.com and chris@riscor.net
Founded in 1912, the SLNHS has remained an active, albeit small society with a core membership of enthusiasts and professionals in nature conservation. It organizes varied programmes of lectures for its members. The subject matter of the talks embraces all fields of natural history, including marine life, birds, environmental issues and the recording thereof via photography and other means. It organizes regular field excursions, which include day trips as well as longer excursions with one or more overnight stays.

Field Ornithology Group of Sri Lanka (FOGSL)

Department of Zoology, University of Colombo, Colombo 3
www.fogsl.net. Email: fogsl@slt.lk
FOGSL is the Sri Lankan representative of Bird Life International, and is pursuing the goal of becoming a leading local organization for bird study and conservation, and carrying the conservation message to the public. It has a programme of site visits and lectures throughout the year, and also publishes the *Malkoha* newsletter and other occasional publications. Education is an important activity, and FOGSL uses school visits, exhibitions, workshops and conferences on bird study and conservation to promote its aims.

Ruk Rakaganno, the Tree Society of Sri Lanka

http://rukrakaganno.wixsite.com/rukrakaganno. Email: rukrakaganno09@gmail.com
Ruk Rakaganno works with communities, particularly of women, to care for and protect water resources and biodiversity. It is also concerned about trees in the urban environment, and currently manages the Popham-IFS Arboretum in Dambulla. Its publications include a newsletter for women in Sinhala, in Tamil and English, and a booklet 'What Tree is That' in the three languages. It also occasionally arranges tree walks.

Wildlife and Nature Protection Society (WNPS)

86 Rajamalwatta Road, Battaramulla
www.wnpssl.org. Email: wnps@sltnet.lk
The WNPS publishes a biannual journal, *Loris* (in English) and *Warana* (in Sinhalese). *Loris* carries a wide variety of articles, ranging from very casual, chatty pieces to poetry and technical articles. The society also has a reasonably stocked library on ecology and natural history. Various publications, including past copies of *Loris*, are on sale at its offices.

The Young Zoologists' Association of Sri Lanka (YZA)

National Zoological Gardens, Dehiwala
www.yzasrilanka.lk. Email: srilankayza@gmail.com
The YZA has nearly 100 school branches and has also set up branch associations. The bulk of its membership consists of schoolchildren and undergraduates, the rest being graduates, professionals and nature lovers from all walks of life.

Forest Department

www.forestdept.gov.lk/web
The Forest Department is the state institution tasked with the conservation and management of many of Sri Lanka's important forest reserves, including the lowland rainforests. It has a long track record of having excellent botanists on its staff, and has produced a small number of publications. It also publishes a scientific journal, *The Forester*. These can be purchased from their head office in Colombo's suburbs.

◾ Acknowledgements ◾

Tour Operators

A strength of Sri Lanka lies in the presence of both general and specialist tour operators that can tailor a wildlife holiday. A non-exhaustive list of companies is given below.

A. Baur & Co. (Travels), www.baurs.com
Arunya Vacations www.aarunyavacations.com
Adventure Birding, www.adventurebirding.lk
Aitken Spence Travels, www.aitkenspencetravels.com
Birding Sri Lanka.com www.birdingsrilanka.com
Bird and Wildlife Team, www.birdandwildlifeteam.com
Birdwing Nature Holidays, www.birdwingnature.com
Eco Team (Mahoora Tented Safaris), www.srilankaecotourism.com
Hemtours (Diethelm Travel Sri Lanka), www.hemtours.com
High Elms Travel, www.highelmstravel.com
Jetwing Eco Holidays, www.jetwingeco.com
Lanka Sportreizen, www.lsr-srilanka.com
Little Adventures, www.littleadventuressrilanka.com
Natural World Explorer www.naturalworldexplorer.com
Nature Trails, www.naturetrails.lk
Quickshaws Tours, www.quickshaws.com
Red Dot, www.reddottours.com
Sri Lanka in Style, www.srilankainstyle.com
Walkers Tours, www.walkerstours.com
Walk with Jith, www.walkwithjith.com

Acknowledgements

DARSHANI SINGHALAGE

My journey with wildflowers began in 1997, when I joined the wildflower study circle of the Youth Exploration Society of Sri Lanka, Royal Botanic Gardens, Peradeniya. Kumudu Amerasinghe, the lecturer in the wildflower circle laid a strong foundation, inspiring me to continue with studying the wildflowers of Sri Lanka, and I am greatly indebted to him. He also helped to identify many species of wildflower during this project, and provided support and encouragement. When I was studying wildflowers with the Youth Exploration Society, I met Professor Siril Wijesundara, who was a superintendent of the Botanic Garden at that time. He understood my interest in flora and made arrangements for me to use the facilities of the National Herbarium. I also thank Mr Upali Dhanasekara, who was a Curator of the National Herbarium when I used it in 1997.

I received lots of input on Sri Lankan flora from Professors Nimal and Savitri Gunatilleke, and Professors Deepthi Yakandawela and Anoma Perera, who taught me different plant-related courses when I was an undergraduate. I thank all of them for the strong foundation. The former director of the Royal Botanic Gardens, Professor Siril Wijesundara, kindly gave us permission to use the facilities in the Royal Botanic Gardens to identify plants, and to photograph wildflowers cultivated in the gardens. We also thank Ranjani Edirisinghe and other staff of the National Herbarium for their assistance with plant identification at the National Herbarium. *A Revised Handbook to the Flora of Ceylon* was a key reference for identification and was consulted repeatedly. I record my thanks to the many authors who contributed to the family monographs in the 16-volume series.

I would like to thank Dr Chandra Embuldeniya, former Vice-chancellor, Professor Rohan Weerasooriya, former Dean, Professor Sisira Ediriweera and all other staff members of Uva Wellassa University for their encouragement and support during this work. My parents, two sisters and Indika aiya and Nadeera's mother provided constant support and encouragement to ensure the success of our work.

NADEERA WEERASINGHE

I would like to repeat my thanks to the people already mentioned in Darshani's acknowledgements. I would also like to thank the members of the Youth Exploration Society of Sri Lanka, who shared their knowledge and encouraged my interests and my wildlife photography. My thanks also to Malathie Irugalrathne, Chamara Irugalrathne, Chamila Wijerathne, Palitha Handunge and Sunil Gunathilake, who supported me in various ways.

I extend my gratitude to all the staff of Jetwing Hotels. A special thank you to Gehan, who has been gently cajoling and encouraging me for more than a decade. Hiran Cooray, Ruan Samarasingha and Gehan all encouraged me to write and photograph when I was a naturalist at Jetwing Hotels. I would also like to record my thanks to the late Upali Weerasinghe and his team at Yala Safari Game Lodge. Tragically, some of my former colleagues, including Upali, lost their lives in the Boxing Day tsunami. This book has its roots in my time at Yala and I remember with respect and fondness the support and encouragement of my late colleagues in Yala. I would like to thank Tyrone David, former training and quality assurance manager at Jetwing Hotels and Renuke Coswatte, former general manager of Jetwing St Andrew's and the other staff of Jetwing St Andrew's. Chaminda Suraweera, a former sous chef at Yala Safari Game Lodge, helped me with my photography. Chandra Jayawardana has been a key mentor, providing advice, literature, and his knowledge and support.

Dr Michael and Nancy Van der Poorten helped us to identify some of the species. Mr Rohan Pethiyagoda helped with technical information and provided encouragement. A big thank you to my mother for her constant support and encouragement to make this book a reality. Finally, I would like to thank Adrian Jayesingha and all the staff of Fire-X and Sir John's Bungalow.

Our thanks to the following photographers for permission to use their images in this book: Nilantha Kodithuwakku (*Vaccinium leschenaultia, Vanda tessellata*), Chamara Irugalrathne (*Pteroceras viridiflorum*) and our daughter Kavithma Nayanadini (*Osbeckia rubicunda*).

GEHAN DE SILVA WIJEYERATNE

Specific Acknowledgements In the early 2000s, I proposed to Nadeera that we begin work on a photographic guide to wild flowers in Yala, where he was working as a naturalist at the Jetwing Yala Safari Game Lodge. I also asked him to work on his botany. He must have taken my suggestion very seriously as he rather helpfully married Darshani, a botanist. My thanks to Nadeera and Darshani for persevering with this book project that has been in gestation for many years. When we began the project, Rohan Pethiyagoda made available to me an Excel database of plant species recorded in Sri Lanka. This has been very useful to us in deriving some of the high-level statistics of species and genera in families in Sri Lanka, and in cataloguing and indexing our images. More generally, I would like to congratulate Rohan for ushering in a renaissance in biodiversity exploration commencing in the late 1990s. He is also to be congratulated for his efforts to give many scientists a voice by publishing their work though the Wildlife Heritage Trust (WHT).

I have a large number of botanical books in my collection that have to one degree or another influenced me. However, there a few publications I need to specifically acknowledge. In writing the family accounts, I drew heavily on *Flowering Plant Families of the World*, by Vernon Heywood, R. K. Brummitt, Alastair Culham and Ole Seberg, published by Firefly Books. A number of other books listed in the bibliography were also consulted for the family accounts. The excellent *The Kew Plant Glossary*, by Hank Beentje, was a constant companion on my desk. Isaack Kehimkaar's *Common Indian Wildflowers* was often in my field bag when I was on field visits in Sri Lanka. For the classification, I used the online work of the Angiosperm Phylogeny Group 4 (APG 4).

The identification of the images we shot was an onerous task and I thank the people who helped us in this. Among these, I would more generally like to thank Siril Wijesundera for over the years being very supportive to a number of researchers and giving them access to the facilities of institutions that he has been in charge of.

Tara Wikramanayake once again assisted greatly with several rounds of preliminary copy editing and posing questions. My thanks again to John Beaufoy and Rosemary Wilkinson for asking me to take on another book, and to Krystyna Mayer for her expert editing.

General Acknowledgements Many people have over the years helped me in one way or another to become better acquainted with the natural history of Sri Lanka. My field work has also been supported by several tourism companies, as well as by state agencies and their staff. To all of them, I am grateful. I must, however, make a special mention of the staff of Jetwing Eco Holidays and its sister companies Jetwing Hotels and Jetwing Travels. Many people at Jetwing Hotels have supported my work. Given the genesis of this book, I would in particular like to record the warm welcome extended by the late Upali Weerasinghe and his team at the Yala Safari Game Lodge. During the Boxing Day tsunami, Nadeera and I tragically lost a few of our colleagues, including Upali Weerasinghe, Viraj Karunarathne and Saman Weerasinghe, as the tsunami took away thousands of lives across the country. Nadeera narrowly escaped with his life and played a big part in the search for survivors and the recovery of the dead. Thanks to my wife's insistence that I spent that Christmas in Colombo, my family and I were spared. Spending several days with Nadeera and other colleagues in Yala recovering the dead made a deep impression on me on the fragility of life and the role of fate, and left me with a renewed sense of mission to do something useful with my life.

My late Uncle Dodwell de Silva took me on Leopard safaris at the age of three and got me interested in birds. My late Aunt Vijitha de Silva and my sister Manouri got me my first cameras. My late parents Lakshmi and Dalton provided a lot of encouragement. My sisters Indira, Manouri, Janani, Rukshan, Dileeni and Yasmin and brother Suraj also encouraged my pursuit of natural history. In the UK, my sister Indira and her family always provided a home when I was bridging islands. Dushy and Marnie Ranetunge also helped me greatly on my return to the UK. My one-time neighbour Azly Nazeem, a group of then schoolboys including Jeevan William, Senaka Jayasuriya and Lester Perera, and my former scout master Mr Lokanathan, were a key influence in my school days.

My development as a writer is owed to many people. Firstly, my mother Lakshmi and more lately various editors including the team at *Lanka Monthly Digest*, *Living*, the *Sunday Times* and *Hi Magazine* who encouraged me to write. Various TV crews, especially the teams led by Asantha Sirimane at Vanguard and YATV, supported my efforts to popularize wildlife. My development as a naturalist has benefited from the programme of events organized by the Wildlife and Nature Protection Society, Field Ornithology Group of Sri Lanka (FOGSL) and the Sri Lanka Natural History Society. In the UK, I have learnt a lot from the field meetings of the London Natural History Society and London Wildlife Trust. I have also been fortunate to continue having good company in the field in Sri Lanka on my visits after I moved back to London. My friends and field companions include Ajith Ratnayaka, Nigel Forbes and Ashan Seneviratne, who have arranged a number of field trips for me.

My wife Nirma and daughters Maya and Amali are part of the team. They put up with me not spending the time they deserve with them because I spend my private time working on the 'next book'. Nirma, at times with help from parents Roland and Neela Silva, takes care of many things, allowing me more time to spend on taking natural history to a wider audience.

The list of people and organizations who have helped or influenced me is too long to mention individually, and the people mentioned here are only representative. My apologies to those whom I have not mentioned by name; your support did matter. Everything I know is what I have learnt from someone else.

Gehan de Silva Wijeyeratne is happy to receive images as additions or replacements for images in his photographic field guides to birds, butterflies and dragonflies, mammals and trees. For more details contact him on gehan.desilva.w@gmail.com